Digitally Daunted

Sean Westcott & Jean Riescher Westcott

Capital Career & Personal Development Series

Capital Books, Inc.
P.O. Box 605
Herndon, Virginia 20172-0605

ISBN 13: 978-1-933102-72-6

Library of Congress Cataloging-in-Publication Data
Westcott, Sean.
Digitally daunted : the consumer's guide to taking control of the technology in your life / Sean Westcott & Jean Riescher Westcott.
p. cm.
ISBN 978-1-933102-72-6 (alk. paper)
1. Household electronics—Popular works. 2. Consumer education—Popular works. 3. Household electronics—Purchasing—Popular works. I. Westcott, Jean Riescher. II. Title.

TK7880.W47 2008
621.38—dc22

2008034117

Printed in the United States of America on acid-free paper that meets the American National Standards Institute Z39-48 Standard.

First Edition

10 9 8 7 6 5 4 3 2 1

To John R. and Jean M. Westcott
—Sean

To the South Dade Regional Library—
for showing me the way forward.
—Jean

To Ian Wolfgang and Megan Grace—
you always inspire us.
—Sean and Jean

Contents

2 Connecting to the Internet

3 Telephones: Landline and Cellular

5 Cameras

Introduction

Too many choices and changes can make life confusing, especially when dealing with technology. Your nephew wants to send you text messages—do you need a new phone? You made the transition from vinyl records to cassettes and then to CDs—how do you get it all to play on the portable music player you received as a gift? Nothing seems easy, and the mistakes can be expensive. How can you get control back over the things that are supposed to make life easier and more enjoyable?

When you try to do your own research, the tech jargon can be overwhelming, and the information usually assumes you know which device you want. Other articles and TV segments tell you what to buy—but you still don't understand why you need it or how it might benefit you.

Technology also evolves quickly. Not too long ago, people would buy three or four separate devices to have the features of today's Smart Phones: a PDA (Personal Digital Assistant), a cell phone, a digital camera, and maybe a personal media player to listen to music. Now you can get all those in one device. But don't feel pressure to keep up to the cutting edge—just focus on the gadgets that deliver what you need and in a way that makes your life a little easier or more fun.

What we aim to do here is to give you a running start. If you understand the basics, know a bit of the **geekspeak**, and learn how to make decisions based on your needs and taste, you will be much better armed to get the equipment or services that actually work for you. To make things easier, each new geekspeak term will be introduced in bold with a short definition in the text and in the glossary section at the end of the book.

But don't let new terms and ideas intimidate you. Remember, each of us knows more than we give ourselves credit for. Think of your life at work. For those of us who went to school as late as the 1980s, we took typing class not keyboarding class. We had to hand type everything into typewriters (no copying, cutting, or pasting), and correcting mistakes required correction fluid, meaning the finished product was often messy and unprofessional. "CC" meant "carbon copy," with actual carbon paper. Can you imagine your work life today if you had to submit things to a secretarial pool, or if you had to operate a mimeograph machine? You probably use word processing software, manage spreadsheets, send e-mails, and manage your schedule on your PDA. Or at least you do one or two of those things. As they used to say, "You've come a long way baby."

The pace of change in workplace technology is probably easier for you to cope with because of the support system of an IT department or at least an IT professional. But incorporating technology into your home and personal life can be relatively easy too.

One of your best resources for good technical advice is your company's **IT** (Information Technology) team. Most IT people are happy to share their expertise, if you remember that you are asking them a favor and if you have done a little bit of homework yourself. The question, "What kind of phone should I get?" is too broad. Try instead, "I want to try using my cell phone for e-mail when I'm away from the office, do you have any suggestions on what phones people are using in our company successfully?" By reading this book and familiarizing yourself with the basics, you will be able to have a more meaningful discussion—respecting both your time and the time of your IT person. Also, remember that a little "thank you" is always appreciated. Organizing a token of gratitude in your office on System Administrator Appreciation Day (last Friday in July; see sysadminday.com for ideas) would be a great way to thank your IT support team.

In our family, having an IT professional—my husband Sean—as part of the "leadership team" has made integrating these changes and myriad devices into our daily lives much easier. As the nontechnical spouse, I—Jean—just want things to work. I don't necessarily need to know how things work to enjoy them. I didn't grow up with computer games, our family did not have a personal computer, and I didn't have a cell phone until we were well into this century.

Sean, however, has always been interested in taking things apart to see how they work. He took apart and rebuilt all sorts of electronic devices. We had Internet access before anyone else in our neighborhood outside of DC, and Sean ran his own website hosted on his own server, just because he wanted to see how it all worked. He introduced me to technology that has made our life a bit more fun and my work more efficient. The technology in our home allows me to work from home at least part of the time, keeping close to our kids' schools. Sean is able to videotape our son's marching band performances and daughter's tae kwon do class and share it with family, friends, and instructors. It allows us to watch the TV shows we like when it's convenient for us as opposed to being locked into the network's schedule—making for evenings with quiet periods for homework and nontechnical entertainment such as reading or playing instruments. And for someone who didn't grow up in an arcade or with computer games beyond Pong and Solitaire, I seem to really spend a lot of time playing Golf on our Wii or rocking out with the family with Rock Band on our Xbox 360.

With *Digitally Daunted*, you will have the advantage of an IT specialist giving you advice, with a nontech person translating that advice in ways you can understand. As a result, you'll learn how to make decisions using good, basic information. The key to using technology is to buy and use the equipment that makes sense and that makes your life easier or more enjoyable. And when you do choose to avail yourself of a new gadget or tool, that you make a decision based on what you want, not what will earn the salesperson their commission.

Remember, you only need to know enough to make a reasonably informed decision. You don't have to know exactly how your new technology operates anymore than you understand exactly how your car works or how an airplane flies. You just need to know enough to select and operate the equipment in a way that gets you where you want to go. And that's what *Digitally Daunted* will help you do, with the least amount of angst and pain possible.

How to Use This Book

Digitally Daunted is designed to provide you with the information you need to become an informed consumer. You may want to read just about the technology you are interested in, or you might want to read the book in chunks. We have included a detailed Table of Contents to help you find the topic you are interested in and the information you need.

We have tried to have each chapter stand alone, defining concepts as they come up. If you come across a word or an acronym that you need a bit more help with, you can refer to the glossary in the back, or to find it introduced more fully, refer to the Table of Contents to see where that topic is first mentioned.

All books about consumer technology reflect what is currently available, and as such, the material requires updates. As this is more of a primer or introductory book, the information is unlikely to require frequent updates, but there will be situations where prices will change significantly or new developments will require that we re-address an application or product. We suggest you visit our website digitallydaunted.com for the latest updates.

Digitallydaunted.com will also feature product reviews, articles, and an area where you can submit questions. Please visit it and let us know what you think of the information contained in this book. We welcome your feedback.

Before We Get Started: What Does "Digital" Mean?

Let's start with a basic concept. What is meant by **digital**?

A bit of history first: before there were digital signals, information was transmitted by **analog** signals. Think of your old television, radio, and telephone. The sound waves (or light waves for pictures) are sent through the air or through wires. These transmissions are very detailed and contain lots of information. However, these signals can pick up interference when they cross paths with other waves of information. This is known as noise. They also take up a lot of **bandwidth**. Bandwidth refers to the amount of space available for information to travel.

As a result of the digital revolution, this same information is now transmitted in a much more compact way—through binary information. All digital signals are composed of a series of zeroes and ones (off and on). Everything you see on your computer—even videos and music—has been rendered down to a long series of off and on switches, or zeroes and ones. As you can imagine, digital video signals require very long strands of zeroes and ones. But with faster and more powerful computers, this information is processed at amazing speeds. Digital signals are also not as vulnerable to noise. In addition, these signals are much more compact, so *many* digital signals can be sent in the same space or bandwidth that used to carry "a few" analog signals.

Don't be intimidated by all of this; be informed—once you grasp the difference between analog (waves) and digital (zeroes and ones), you can understand why everything around you is going digital. And remember that "grasping" the concept doesn't mean you have to pass a test on it—all you need to know is the basics.

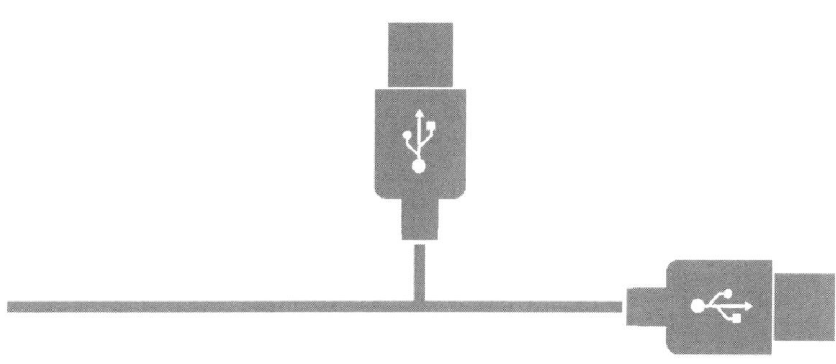

1 Your Computer

Choosing the type of computer you want involves thinking about what you want to do with it. All home computers will help you balance your checkbook or print a shopping list, check your e-mail or view web content, but there is so much more you can do: edit photos or home movies, play extremely realistic games with hundreds of other people online, store and organize a music collection and even broadcast it wirelessly so you can listen to it on different devices around the house, or download and watch new movies on your television or computer monitor. Some computers will perform all these things, and others will only do a few of them. You will have to decide what you want and weigh that against how much money you want to spend.

Let's jump right to the basics—we'll cover the more advanced options later in this chapter.

A Word on Budget and Prices

The first thing you have to accept is no matter what computer you buy today, there will be a cheaper and better computer out in a month. Whatever is labeled "cutting edge" today will be widely available in the near future. That's frustrating, but there's nothing you can do about it. So set your budget by what you need and want in a computer right now, and look at the range of prices currently available.

The fast-changing nature of technology also means that after a few years of reliable use, you might want to consider replacing your home computer instead of repairing it.

Geekspeak—Computers 101

Following our *Digitally Daunted* philosophy, you don't have to know *everything* about a computer—how it's built or how it works—but you do need to know a few terms so you can get the computer that you want and need.

PC is the generic term for a personal computer. Most PCs run the Microsoft operating system (Windows 98, XP, Vista, and others). **Macs** generally refer to computers manufactured by just one company—Apple. Apple both produces the **hardware** (the actual parts of the computer), the **operating system (OS)**, and most of the **software** (the programs that run the computer). Apple computers, therefore, do restrict the add-on software and hardware options available to you, but many users appreciate and favor their relative ease of use and attractive presentation.

A **Central Processing Unit (CPU)** or **processor** is the brain of your computer. The software that operates it is the **firmware**. What you think of as your computer is usually the case or tower containing the CPU on a **motherboard** (the circuit board that holds the main electronics of your computer), plus a fan and other electronic circuits. In Apple computers, the motherboard is sometimes called a "logicboard."

An operating system (OS) controls everything from the programs to the hardware and how they work together. It is the "graphical user interface" you may have heard about. This means that instead of looking at a lone blinking cursor on a black screen—you may remember these old Disk Operating Systems (DOS) in which you had to type special codes to do anything—you can now click on user-friendly icons and buttons to perform various tasks. The development of easy-to-use operating systems has taken computers from being strictly useful to science and business to being essential tools for everyday life. In other words, user-friendly operating systems make computing possible for the everyday person.

Microsoft Windows is the most popular operating system. The various Apple operating systems are used by a smaller but significant percentage of Apple computer users. Another operating system you may have heard of is Linux, which is an **open source program** (a program developed collaboratively and available for free for most personal uses).

Memory or **RAM** is the storage function on your computer. It is used to store files and programs, to manage information that needs to be accessed quickly, and to retrieve files stored on the hard drive. Memory size is usually expressed in **gigabytes (GB)** or **megabytes (MB)**. (See the "Geekspeak" box in this chapter for more information on gigabits and megabits.)

A **hard drive** is a storage device that retains your

information even when it is turned off.

There are other components basic to your computer, such as CD and DVD drives. Most computer setups will include other fairly standard **peripherals**, which are add-on items such as a keyboard, mouse, printer, external hard drives, and the like.

Computer Shopping—The Basics

Shopping for a computer can be stressful and overwhelming. It is an important tool that you will use daily, and it is a significant investment. You want to find a computer that matches your current and projected needs for the next few years. If you break the purchase down into several key components, it will be easier for you to buy your next computer with confidence and to choose a system that works for you.

Monitor

You spend a lot of time looking at the computer screen, so a nice, big, easy-to-read screen on a **monitor** is important. Beware, though, you can blow the budget really quick getting the biggest and fanciest monitor. Upgrades in this department can add up fast, so ask yourself: do you really need a twenty-four-inch monitor to read e-mail?

A seventeen- or nineteen-inch screen will work great for almost everything. If you plan on playing high-end video games or making movies, or if it is a really good sale, then go for the deluxe-size monitor. The good news here is that a nice screen is often the first thing vendors put on sale to get you to buy a computer.

There are two basic types of monitors available. **CRT monitors** are the large and usually beige plastic monitors that you may have used for years. These are quickly becoming obsolete in favor of **LCD** (Liquid Crystal Display) or other screen monitors. If you are buying a new computer (that is a new computer processing unit), you do not need to upgrade your existing monitor. But if you still have a CRT monitor, it makes sense to purchase an LCD monitor when you're upgrading your system, as the package pricing is usually better than purchasing a monitor separately at a later time.

Geekspeak

You hear these terms all the time, but you might not be sure which is bigger—a megabyte MB or kilobyte KB. Here is a basic chart:

1 Byte (B) = 8 bits; the size of a single character

1 Kilobyte (KB) = approximately 1,000 bytes; about the size of a paragraph of text

1 Megabyte (MB) = approximately 1 million bytes or 1,000KBs; a song purchased from an online store in the MP3 format is about 5-8MBs. The old floppy disks generally held around 1.4MBs.

1 Gigabyte (GB) = approximately 1 billion bytes or 1,000MBs. You can store about twelve hours of music in MP3 format in 1GB or about 1,000 novels.

Memory

It's worth spending money to get the most Memory (RAM) available. More RAM means more files and programs can be open at one time and big files can be accessed quickly. Almost all new systems come with at least 1GB. But if the budget allows, go for 2GB, at the least, and as much as the system can take. This will make your computing experience much better. Memory upgrades are currently around $50 to add 1 GB and $100 to add 2 GB.

Hard-Drive Space

A hard drive is where you will store most of your files, i.e., programs, documents, music, photos, and videos. Files take up space. An average photo is around 1MB, and a small document is half that size. A program to open a file, photo, or document takes up around 500MB to 1GB. For a laptop, 150GB of hard-drive space is the minimum we would recommend, and for a desktop, 500GB.

Processor

As we have said before, the processor is the brain of your computer. You may be wondering why the brain is third on our list. It's because all modern computer processors or CPUs will get the job done and get it done fast. One thing to look for if you want that extra power is something called **dual core** or a quad core. This means that there are two (for dual core) or four (for quad core) CPUs in one, so more work can be done faster. Almost all newer systems come with one of these types of CPUs, but you should make sure it does.

Note: These four items—the monitor, memory, hard-drive space, and processor—are the things we recommend spending your hard-earned money on. Upgrading these will greatly increase your experience. On the other hand, you can buy a $500 laptop or PC, upgrade nothing, and then use the money saved for a nice printer or a digital camera. The upgrades mentioned here are just suggestions. All but the CPU can be easily upgraded later if you find you need a more powerful computer or a bigger monitor.

Working from Home

If you will be using your computer to work from home and connect to a work network, you should definitely speak with your IT department before buying your next computer. Ask about any special considerations or requirements necessary to log onto the network. Do you need specialized software? Is there a specific operating system you need? Do you need any special **peripherals** (add-on items such as printer, mouse, etc.)? Also, you will need to alert them so that you can set up access to the work network. A phone call in advance letting them know you're buying a new computer can save you from waiting days to be able to get back onto your work network.

Other Purchasing and Upgrade Options

All PCs and laptops come with other options that usually don't need to be upgraded, but you should know their function and what to look for.

CD/DVD Drives

CD and DVD drives are usually called **optical storage devices** (as opposed to magnetic storage like your hard drive or the older storage media like floppy disks or cassette storage). With these drives, data is stored on a disk that is read by a laser.

There are a couple of choices here. Using a **CD writer**, you can save data, music, photos, or videos to a removable disk; this is called "**writing to a disk**" or sometimes called "**burning to a disk**." Most systems now have a combo drive that reads both CDs and DVDs. Most systems also allow you to save files to blank disks, so you can make a CD of music to listen to in the car or make a home DVD to watch with your friends.

Most new PCs come with at least a CD writer, and many come with a DVD writer. There are different formats of DVDs however. They are indicated by the letters following DVD. They are "-R" and "+R" and "-RW" and "+RW." The "-R" and "+R" are two different types of recording, and the "W" indicates if the disk is "**re-writeable**" (able to be used more than once to record).

Usually the two formats, "-R" and "+R," are interchangeable, with one significant difference. Some older DVD players will only play the DVD-R formatted disks. *To be on the safe side*, look for a DVD writer that can record in both formats. You may also see "**dual layer disks**." Regular DVDs can usually hold up to 4.7GB of data; dual layer disks can hold up to 8.5GB.

Video Card

When Microsoft introduced its new operating system,

Vista, most PC and laptop manufacturers upgraded the standard **video card** included in their products. (A video card is a piece of computer hardware that contains the electronics that create the display on your computer's monitor). To ensure that a video card will work well with current software, make sure there is at least 256MB of dedicated memory on the video card. This means the video memory is for the video system only and not shared with the rest of the PC.

Sound Card

All modern PCs have a **sound card** (the electronics that produce the sounds of both the computer—"You've got mail"—and the videos or music played through your computer). In addition, almost all support surround sound (see Chapter Four, "Television," on page 69). That way you don't have to consider a separate sound system as an option to add on to your computer. You will still need **speakers** to listen to the sound, though, as the internal speaker will not produce quality sound replay.

Speakers for computers can range from the really inexpensive (most computer packages include at least basic stereo speakers) to complete surround-sound, five-speaker systems. This is a matter of choice and how you plan on using your computer. We recommend going into a store and listening to various speakers and to choose one that works with your budget and sounds good to you.

You may also want to invest in a comfortable set of headphones as well. If you haven't tried using over-the-ear **headphones** as a way of listening to music and movies on your computer, you may want to consider how they can help reduce distraction from outside noises or to allow you to listen to audio without disturbing others.

USB and FireWire Ports

Something a lot of people do not consider is the number of ports available on their computer. A **port** is an "interface" on a computer, or simply put, the point of contact where you connect one device to another. Think of an electrical outlet in a wall and a plug—the port is the electrical outlet. You want to make sure there are enough ports to connect all of your devices, including things such as your digital camera and personal media players.

There are different types of ports. Some have dedicated purposes, such as your keyboard port, your printer port, your monitor port, your keyboard port, or your mouse port. These types of connectors are quickly being replaced on newer PCs by other ports that are more flexible in their use. Of these, the most popular is the **USB** (Universal Serial Bus).

A USB port allows you to attach external devices to your PC, such as a printer, scanner, external hard drive, your camera, and other items. You will see the image (Figure 1)

Figure 1:
USB or Universal
Serial Bus

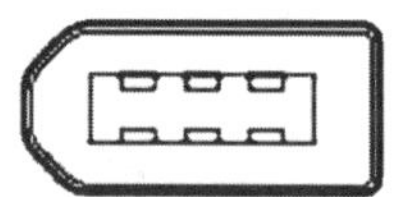

Figure 2:
Two types of FireWire
Ports (4 pin and 6 pin)

imprinted next to the USB port on your computer.

Ideally, you want to have a lot of USB ports on your computer. We recommend four or more *on the front and back* as it's no fun climbing behind your PC (especially if it's against a wall or under a desk) to plug and unplug devices.

A FireWire port (see Figure 2) is similar to a USB port, but they are not as popular anymore. However, most high-end video cameras still use the FireWire port to transfer video to your PC. The FireWire port can also be used to control the camera functions, such as rewind, stop, play, and fast-forward.

Video-Related Upgrades

We have suggested some basic upgrades thus far, but what if you have a specific task in mind, such as video editing? Even if you know nothing about editing video files, you may want to experiment with it down the road, as near-professional quality videos are now creatable by almost any computer user once they learn a few skills. For instance, you may have a teenager in the house who wants to create videos for school projects or you may want to create a short video to show at a family gathering. In that case, it's better to get the necessary tools and upgrades when purchasing your computer system so that you can avoid expensive after-the-fact upgrades.

Video files are extremely large. For example, a three-minute HD video can range from 200MBs to tens of GBs depending on the quality settings, so all the upgrades I mentioned previously in this chapter will be needed if you want to edit videos. You will probably need to go for even higher numbers in terms of more RAM, a bigger hard drive, and a bigger monitor. While it's true that, to some degree, any computer will edit video, it can become tedious to work on a system that is not up to the task.

Gaming Upgrades

Chances are that if you feel "digitally daunted," you are not what people call a "**gamer**." Nevertheless, online or computer games are one of the most popular things to do on a computer. When we refer to "**gaming**," we are generally referring to simulation-type games or interactive games based in virtual worlds. If you aren't a gamer now, don't assume that you would not ever enjoy exploring online realms or piloting a virtual fighter jet.

If you want your computer to be able to play these increasingly sophisticated games, you may need to buy a more powerful computer. That means a faster processor, more memory, video cards with more memory, monitors capable of providing the most realistic images, and even a specialized keyboard and mouse. These systems can quickly become expensive but they deliver an experience that is truly awe-inspiring.

Media PC

You should also consider whether you need or want a **media PC**. This is a computer that you can connect to your stereo system and TV. A media PC can perform all the tasks of a normal computer, but it can also store video files, music files, and photos in an integrated way, with special connectors so these things can also be viewed on your TV. See Chapter Four, "Television," for detailed information on media PCs, as you may want to consider this as an option when purchasing a computer.

Extended Service Plans

One option you may consider is the retailer or manufacturer's extended service plan (it may be called other things such as a "buyer's protection plan" or "extended warranty"). In general, we do not recommend them. If you need a repair, it will most likely occur during the normal warranty period. If the problem arises after the warranty period, the cost to fix the problem will be around the same as the cost of the service plan. In addition, most of these plans do not cover accidental damage caused by you, such as dropping the computer or spilling coffee on it.

Don't overlook the warranty provided by the manufacturer of your computer and its components and peripherals (the added-on items such as the mouse, printer, scanner, etc). Most provide a period of free phone support and free repair of defective hardware. Once you are outside of the warranty period, you may also be able to pay á la carte for phone support. All of the major PC manufacturers and Apple will allow you to extend your warranty during the warranty period. If you find that you use the phone service and are happy with the help provided, you might want to extend the service.

Laptops are a special case when looking at a warranty or extended-service plan. Think about how you will use your laptop. If your laptop will mostly be used in an office or home, without a lot of travel, you may run into a situation where you need more than the basic warranty. If your laptop will be on the go, in precarious situations such as on a construction site, or if you have a tendency to drop things, you might run a higher risk of damaging the screen of the laptop. This very common situation will have you facing a very expensive repair (sometimes more than half of the cost of the laptop!).

Most extended warranties do not cover accidental damage, theft, or loss. In this case, you may want to consider purchasing laptop insurance instead. While most of your personal property, including your computer, may be covered by your homeowner's or rental insurance, you might want to carry a separate policy for your electronics. Consult your insurance agent for their advice on this matter. Make sure you get replacement value instead of a depreciated amount. You may be surprised to see how affordable this insurance can be (less than $75 a year).

Retailers' service plans are sold as add-ons to your purchase, usually at the time of purchase. Some will allow you to sign up for the extended warranty during the return period for the product. These extended plans are high-profit items for the retailers but may not be the best choice for you. Don't allow the pressure to force your decision. Most manufacturers will allow you to extend their warranties directly within the initial warranty period. In addition, if you do need a repair, retailer service providers are usually offsite, which means that you may be without your computer for up to two weeks if it has to be sent out for service.

Setting Up Your Computer

Bringing home a nice, new computer and setting it up

and getting it running may seem like a daunting task the first time you do it, and you may get a lot of sales pressure to pay to have it set up for you—but trust me you can do it. All computer manufacturers want you to have a good experience with their products and include step-by-step instructions with photos that are easy to understand. It also is valuable to know how your computer is set up when you encounter problems or want to add devices. It can be a great boost to your confidence when you realize that you really can do it yourself.

Most of the cables and plugs are going to be color-coded—meaning one end of the plug will have a color that matches the port where it should be plugged in. Non-colored cables will only go in one place and are keyed to fit in only one place; an example of this is the power plug. It will have only one place where you can put it in and it will only go in one way. If in doubt, don't force a cable or plug in. All the connections should go together easily with little or no pressure. Sometimes the keyboard and mouse will have USB connections, so there are usually several slots they can go into. It does not matter which one you use.

After you get all the cables connected, you can turn everything on. You will get a welcome message and the system will walk you through setting the rest up. It will ask you some questions like your name. Always keep in mind many people have put a lot of time into making this as easy a process as possible. Even the answers to these set-up questions can be changed later. All include a phone number to call if you encounter any problems with setup.

If you bought a printer or a camera with your

computer, you should hold off on setting them up until after getting the computer itself up and running. When you add these peripherals, be sure to read the manual to see if you need to follow a particular order in setting up the devices.

Of course, you may want some help migrating (which simply means moving) your files from your old computer to your new one. Or you still may choose to hire someone to help you set up your system. We simply encourage you to try going solo and not to be *daunted* by your new purchase.

Where to Buy and How to Buy Your Computer

You have lots of options when it comes to where and how to buy your computer. As with all of these decisions, we recommend asking people you know and trust for their recommendations.

Never overlook the resource you may have right at hand—your company's IT department. If you have a good relationship with your corporate IT person, they can be an invaluable person to know and to consult for advice. Be polite, but don't be shy about asking their opinions on where to shop or any pointers they have. Also, ask your friends and colleagues, be specific: Where did they buy their computer? Are they happy with their computer? Was the purchase process easy and without "glitches"? Did they feel the salesperson addressed their questions?

Remember, you are in control of your shopping experience. If you cannot shop without feeling pressure to buy options you don't want, or to buy right away without exploring your options or comparing prices—do not give that vendor your business. In-person shopping gives you the opportunity to see, touch, and use the computers.

Online retailers have helpful questionnaires that help you build a custom system. We recommend giving each a "test-drive"—go through the process of choosing a computer without buying that day, and then compare the two experiences.

If you choose the in-person route, don't assume the national electronic chains are your only choice. There are many independent dealers that can offer valuable advice and service. If they have survived the big-box retailing competition, odds are that they are doing something right. You can also use your warehouse club if you are comfortable with no hand-holding at the time of purchase.

If buying online, you can use the websites of the national electronics chains or warehouse clubs. There are also some online-only electronics stores. Many of those provide the most competitive pricing but often cater to IT-savvy customers. The most popular online shopping is manufacturer direct. Often the pricing here is very good as well, and there is a lot of self-help in the way of customer service and automated tools.

Buying used computers is sometimes the most difficult and unpredictable experience. Unless you can get an exceptionally good deal, we would suggest extreme wariness. You want to be sure that the computer is not stolen and that you receive some guarantee that it's not defective.

Checklists

Before You Shop for Your Computer

What is your budget range? ______________________

List what you use your current computer for:

____

____

____

____

What don't you like about your current computer?

____

____

____

____

What new things would you like to use your computer for?

____

____

____

____

If applicable, have you discussed any work/home computer compatibility issues with your work-related IT specialist? Y/N

Do you want or need a new monitor? Y/N

Do you want or need a new printer? Y/N

Comparison and Price Chart

Type of Computer			
Monitor size			
Operating system			
Hard drive memory			
CPU type			
Optical storage (CD/DVD drives)			
Video card			
Speakers			
Number and type of ports			
Price			
Additional features:			

Computer Information Records

Manufacturer and model number: ____________________

____

Serial number and date of purchase: __________________

____

Warranty term: _____________________________________

____

Operating system key number (located in the original packaging):

Additional items under warranty (monitor, video or sound cards, printer, scanner, etc.):

____

____

____

____

____

____

Did you register your devices? Y/N

Did you register your software? Y/N

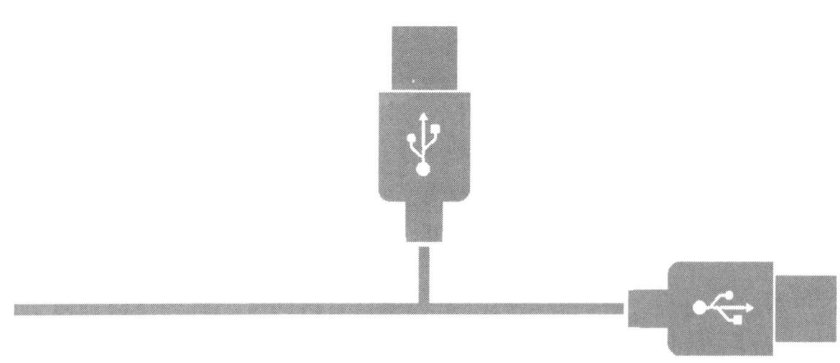

2 Connecting to the Internet

Most of the great leaps in office efficiency are due to the Internet. It has transformed how we do almost everything in our workday. For many of us, the Internet at work is a fast and seemingly limitless resource. So why should we deal with slow or unreliable connections at home?

Dial-up used to be the only affordable or even available way to log on to the Internet. Now we have choices. The increasing availability of high-speed Internet makes it possible for you to add new entertainment and information options to your home: movies on demand; almost unlimited music libraries; and photo sharing, movies, and video phones for connecting with friends and family scattered all over the world. With high-speed Internet connections, these features are now accessible in speeds that are light years faster than the first days of dialing up on a shared home phone line.

Decision Keys

The decision on how to connect comes down to these four Decision Keys:

1. **Availability**
2. **Speed (Upload vs. Download)**
3. **Reliability**
4. **Cost**

In most locations, you may see a lot of advertisements for various types of high-speed Internet connections, all claiming to be the fastest. In short, dial-up is slow, and the **broadband** options are fast. Broadband is anything that is not dial-up—that is, anything that is faster than a modem over a telephone line. Which broadband provider you choose often is complicated by local availability and pricing that is changing quickly as the networks build themselves into neighborhoods.

So, before we get started with the pros and cons of each type of Internet connection, you should learn a few words of geekspeak:

Geekspeak—Internet 101

Bandwidth: You will hear this term tossed around a lot. It refers to the speed of your connection. Connection speed is expressed in bps (bits per second, usually there are 8 bits in a byte; Kbps = Kilo bits per second; Mbps = Mega bits per second; Mbps is much faster). Typical speeds are as follows: maximum speed for dial-up is 56Kbps; DSL speeds start at 720Kbps, and cable Internet speeds are typically around 1 to 2Mbps. What you need to know: more Bandwidth is what you want.

Upload: When you send an e-mail, you are uploading it to another person or network. When you put a file on your computer at work, you are uploading it. Upload speed is important if you plan on playing online video games or transferring large files such as photos or videos to other people or to web-sharing sites like Flicker.com or Google's Picasa.

Download: This is the opposite of upload. When you visit a website and open the homepage or when you access your e-mail, you are downloading those items onto your PC.

Modem: This is the device used for connecting to another network—the Internet is another network, just a really big one. You can also connect to your work network via a modem. In the past, a modem referred solely to the part of your PC that you plugged your phone line into, but now it can also mean a "cable modem," which is a box that connects to cable Internet, or a "DSL modem," which connects to **DSL** through your phone company. Your cable or phone company will usually provide a cable or DSL modem when you subscribe to their service.

Network Cards: A wireless network card gives you the ability to connect your computer to various networks—such as the Internet or your office network—without wires. In other words, you don't have to plug into anything. You can take your laptop onto the porch and surf the Internet from there. You can also use a "wired network card," sometimes referred to as an Ethernet card. This uses a connecting cable that looks a lot like a phone cable, except that it's bigger. Network cards transfer data much faster than a modem.

Routers: This is a device that connects one network to another and directs the traffic between them—not to be confused with a firewall. (More on these in the "Home Networking" section, later in this chapter.)

Firewall: This is a hardware device or piece of software that protects your PC or network from unwanted access. For most home use, you will want to get a combination firewall and router. (More on these in the "Home Networking" section, later in this chapter.)

Methods of Connecting to the Internet

Dial-Up

Dial-up Internet access uses your telephone and a modem to connect your computer to the Internet. Usually there is a toll-free access number, and the service provided is fairly bare bones and reliable. However, the Internet is increasingly being designed with fast connection speeds in mind. By using a slow connection such as dial-up, you are losing or missing out on the amazing functions and capabilities of both the Internet and your computer. If you have dial-up, you'll find that you cannot access some sites that you would like to access (or the sites take forever to download), and you may have trouble sending large files or photos to people.

Nevertheless, many people still use this access method for its affordability and then "make do" with slow surfing on the Internet. Others do not have access to faster connection methods due to their geographic location. As the Internet becomes more ingrained in our daily lives, to the point that many government functions and educational opportunities become unavailable to those without fast connections, this will become an even larger issue in local, state, and national politics.

Also, because dial-up uses your home phone line, you

need to either choose between your phone's availability (callers will get a busy signal when you're online via a dial-up connection) or invest in a separate dedicated phone line. Adding a separate phone line may erase any cost savings that dial-up holds.

There is one advantage to dial-up: when you travel, you may be able to dial-up the Internet from any location that has a phone line, say a vacation home that is remotely located and that doesn't have access to other types of connections—wireless, cable, or DSL. But be aware of any long-distance charges that may result from dialing into the Internet from a new location.

Decision Keys

Availability: Widespread.
Speed: Very slow, limiting; max 56 Kbps upload and download.
Reliability: Good reliability.
Cost: Low cost compared to fiber optic or satellite.

Cable

You may already have cable TV in your home, and if so, then high-speed Internet from your cable TV provider may be a great option with little to no changes in your home. This is the most common type of high-speed connection available in the U.S. It is relatively affordable and extremely reliable if you have a good relationship with your cable TV provider. If your cable TV is constantly going out or if the picture is often bad and customer support is no help, then the Internet service you receive is likely to have similar problems. However, if your TV is almost always available with a good picture, and if cable is restored quickly after a storm outage, and customer support is responsive and helpful, then you should have a great experience with a cable connection.

One thing to keep in mind here—even though a big company may officially be providing your cable, the service is often managed and maintained locally, and each local service is different. Big company "C" in one city or town may provide great service or extremely bad service in another city or town. So ask your neighbors if they use this service and if they are happy with it.

As cable doesn't use your phone line, no separate phone line is necessary. Cable Internet doesn't interfere with your cable TV.

Decision Keys

Availability: Available in most areas that have cable TV; check with your local cable company.

Speed: Extremely fast. Up to 30Mbps download speeds; upload speeds vary greatly but are much slower around 1 to 2Mbps, usually lower than 1Mbps.

Reliability: Good, but varies by local network; customer service is spotty, ranging from very good to extremely bad.

Cost: Generally $40 to $80 per month; but bundled pricing with cable TV and/or phone service is usually offered.

DSL

This is the second most popular type of high-speed connection available in the U.S. and gaining quickly on cable. We believe this type of connection will likely be replaced in most areas with newer technology, such as **fiber optic services**, but generally that is still some years away.

DSL is provided by your local phone company or it can be purchased from a third party. A third-party provider will still be using your phone company's cables and network to get it to your house. This is important because if you choose a third-party provider and something goes wrong, you will be dealing with two companies to get it working again.

Another thing to keep in mind with DSL is the distance from your home to the phone company's **switch office** (network connection) location. The further away you are, the slower your connection speed. If you are too far away, DSL may not be an option at all. Because DSL is a high-frequency electrical signal carried over copper phone lines, the physical properties of the copper limit the distance the signal can travel without distortion or signal strength loss. Check with your DSL provider. A reputable supplier will guarantee performance and allow you to back out of the contract if service is spotty from the outset.

No separate phone line is required, but landline service is necessary.

Decision Keys

Availability: Widely available, usually clustered in urban and suburban areas.

Speed: Fast connection—up to 10Mbps download, but as with cable, you will have much slower upload speeds, usually less than 1Mbps.

Reliability: Good reliability, but distance to phone office switch can make this a poor choice.

Cost: Prices generally start at about $35 per month, but can be significantly higher.

Fiber Optic

If this is available in your area, then this is a great way to go for the most modern Internet experience. Be prepared for several installation appointments. Because this is new technology, there may also be some road bumps before it's all working smoothly. Once it is installed and running, though, you should have smooth sailing with the fastest connection you can get with lots of great features.

You can also get your phone, your TV, and your Internet all from one source using a fiber optic service. The installation usually includes building a home network for your computer and TV so they will all "play nice together," and customer support can help with all aspects of your new digital home.

This is an integrated service providing phone, TV programming, and Internet.

Decision Keys

Availability: Not widely available; usually clustered in urban and suburban areas.

Speed: The fastest, up to 100Mpbs upload and download.
Reliability: Once established, very reliable.
Cost: Tiered pricing dependent on speed and features; starting around $45 per month.

Satellite

This can be a great choice if you have no other high-speed options where you live, as satellite service is available everywhere in the continental U.S. It offers fast downloads, but uploads are quite slow. This means that you can't play online games that require fast action, but for viewing web pages and downloading music or video, satellite service does the job.

This is completely separate from your local phone service.

Decision Keys

Availability: Available everywhere.
Speed: Fast up to 1.5Mbps download; 128Kbps upload.
Reliability: Reliable, but may be affected by atmospheric conditions.
Cost: Very expensive; requires professional installation paid upfront (usually around $300 after rebates) and monthly fees of about $50 to $80; an annual contract generally is required.

Free Community Wi-Fi

A lot of communities are offering free or inexpensive

wireless Internet access. This can be a good option if you live in such an area, but I would not recommend this as your primary way to connect to the Internet. Consider this to be a nice option for working outside or in a café.

WISPs, WiMax, and Others

Wireless Internet Service Providers (WISPs) are one of the newer ways to get high-speed Internet, and all the details are still very much in flux. The main providers of these new technologies are the cell phone companies, but the cable providers will soon be getting involved, as well as some major software providers like Google and Microsoft.

This is all very new, but for updates you can go to digitallydaunted.com, where we will be updating information on new technology as it evolves and answering your questions.

Home Networking

Even if you plan on having only one PC or laptop use a high-speed Internet connection (this section is not for dial-up connections), it is important to have a good home network in place. You never want to connect directly to the DSL, cable or fiber modem without some kind of device protecting you. A combination router/firewall is highly recommended. Most Internet providers will provide the equipment and set it up for you; there will be an added fee and most likely a small increase in your monthly bill.

You may even have a network in your home and not even know it.

Setting it up yourself does require a little bit of preparation but it can be done by even the most daunted. Home networking equipment is designed for the novice, and the instructions and installation CDs will walk you through the set up. Most often, the only necessary equipment is called a **broadband router** or a **cable/DSL router** and will allow you to connect more than one PC or laptop to it. These devices will also have a built in firewall. A **firewall** is a device that inspects incoming connections to your computer network and following a set of rules either allows access or denies it. A firewall protects any device you add to your home network be it a PC, laptop, or media server. It is not important to completely understand all that a firewall does but if you have a high-speed connection you NEED a firewall. Unprotected computers connected to the Internet have been shown to be attacked within twenty minutes of their first connection.

Going Wireless

You may want to make your network a wireless one. A **wireless network** allows you to connect to the Internet and the rest of your network without using wires or plugs. Instead of a standard router, you will want one that includes wireless capability. This broadcasts your Internet signal to your devices so that you can use your laptop on the couch, at your kitchen table, or anywhere within range. To set up your network you need to follow the directions included with your wireless router carefully. Two important steps that are often overlooked are renaming your network and changing your password (which is usually called a **SSID**—service set identifier). If you don't perform these two steps you may be granting access to your network to anyone within its range. This may seem like a harmless thing, but remember that you may be responsible for any illegal activity conducted over your network, and by sharing your connection, you reduce the performance of your network. Each of your computers and other devices that you connect to your network will need to know the name of your network and the SSID/password.

When you take your portable devices out and about, you can often connect to the Internet via **Wi-Fi** (pronounced "wye-fye"; Wi-Fi is a trade name for a technology that allows you to connect wirelessly to the Internet). Many shops, libraries, airports, convention centers, or even local governments provide a shared wireless connection. There are both paid and free wireless networks. Remember that this is not your home network and practice safe computing.

Privacy and Security

Basic Internet security protects your hardware, your Internet connection, and most importantly, your privacy. As in any community, there are people dedicated to wreaking havoc, attacking others with hidden identities, and stealing money. Just as you lock your house and car and don't leave your financial data lying around in public places—you need to be vigilant with every aspect of your computing.

You may have access to a world of information and people via the Internet, but that also means that the entire world has access to your computer and the information you store on it if you do not take basic precautions.

The attacks come in several forms:

- Using your computer to SPAM other computers.
- Using your computer to attack other computers or spread viruses.
- Accessing your personal or financial information and selling it to identity thieves.
- Disabling or damaging your hardware.
- Using your computer to illegally download or upload copyrighted or illegal materials such as certain types of pornography.

As you can see, you don't want your PC to be subject to these types of attacks. At your workplace, there is a dedicated person or department to ensure your company's information and networks. So what can you do to protect your PC and home network?

Install Security Software

Your **Internet Service Provider** (ISP, the company that connects you to the Internet) will usually offer a software security suite. This usually includes free anti-virus software, basic firewall protection, auto-updates, and follow-up and support when you suspect an attack. **Anti-virus software** protects you from computer viruses and worms that attack your system to either damage your computer, steal your private information, or send out SPAM or viruses to other computers. Beyond protecting your computer from an attack, you also want to protect yourself from spyware. **Spyware** is software that you can inadvertently install on your computer that monitors your computer Internet browsing and searching without your formal consent. Common ways of getting infected with spyware is by visiting sites that offer performance upgrades, animations to use on your e-mail or chats, or that offer free security scans. Spyware can slow your computer, damage files or make some of your software inoperable. Your ISP provided security suite might include an **anti-spyware software** tool. Be sure that with both your anti-virus and anti-spyware software, you enable the auto-updates feature so that your computer is protected from the constantly evolving threats.

Update Your Software

Updating your software is essential. Most modern software and computer operating systems have basic protections, but the game is always on to find ways past those measures as hackers look for holes in the security. It is important that you have your computer software obtain the updates offered by the software manufacturer. Many of these updates are security related, and once a known security hole exists, if you fail to update, you are an easy target.

When you first set up your computer or install new software, you should turn on the **auto-update** option. If you choose not to enable auto-update, make sure that you check for updates on a regular schedule (once a week for frequently used software, or on next use for something that you use irregularly).

Practice Safe Computing

Basically, you need to be hyper-vigilant and suspicious.

A good rule of thumb is to apply the approach that you would to eating food. Would you eat food that you received for free from a stranger? No, you would only accept food from a trusted source such as a restaurant that is inspected by the health department or from a good friend or family member.

Do not open attachments from an unknown source. Be aware that e-mail addresses can be "spoofed" or copied to look like they were from someone else. Would you open a package of cookies left on your front door with a note, "From Mom." You would most likely use common sense to determine if your mother really did drop off cookies. You would probably call and check. Likewise, don't click on, open, or run anything that "smells funny."

Pay Attention to Performance Problems

Don't ignore performance problems. These might be a symptom of an attack or of impending hardware failure. Take notes on any error messages that appear on your screen, problems encountered, and what you were doing at the time of the problem. See Chapter Seven for more information.

Avoid Illegal Computer Use

Be aware that you are responsible for what is done on your computer by you and your guests. This is especially important in two major areas: copyright violation and pornographic material. The biggest danger is that you will be criminally responsible, as possession of some materials (such as child pornography) is a crime in itself.

You may also be found to be financially liable for

violating the copyrights of music and movie producers. Talk to your children and your guests and be sure that they are using the Internet responsibly. One of the easiest ways to install viruses and other "malware" is by visiting sites that purport to offer "free porn" or "free music," or by clicking on e-mail attachments that offer those things. Hackers and thieves know human weaknesses and are happy to exploit them.

Back-Up Your Data

Mistakes, attacks, power failures, and hardware failures happen. Don't lose everything, including those irreplaceable items that you only have digital copies of, like photos or financial information.

You can create a **back-up** copy of your data—this is called "backing up"—which allows you to recover anything lost due to damage. Basically, you want to create either a physical copy either by putting it all on DVDs or on an external hard drive, or by using a web-based backup service. See the "Disaster Prep" section of Chapter Seven.

Be Careful Computing on the Road

When you are away from home, remember that you don't have all of your home protections. The Internet connection you're using on your travels may be compromised. A good rule of thumb is to view any "new" connection as public-forum computing. Or compare it to speaking on your cell phone in public. As much as you can, try to avoid visiting financial sites or accessing private information. This is especially true when using Internet terminals. These may have been compromised by "key-loggers" that record keystrokes to reveal personal information such as passwords. It is best to be vigilant when using these while traveling, as it is so much easier to prevent problems than to correct potentially costly ones.

Checklists

Internet Service Providers Comparison Chart

Name of ISP			
Speed (upload)			
Speed (download)			
Price per month			
Modem rental fee (Y/N); cost			
Contract Terms			
Comments from Neighbors/ Friends			

Internet Service Provider Records

Fill out the following information, and keep it handy for service calls—you will often be asked for this information. *Tip*: Fill this out at time of purchase/installation.

The name of your Internet Service Provider:

____

ISP account number: ______________________________

ISP customer service phone number: _______________

What is the name of your browser?: ________________

Modem manufacturer/model number: _______________

Modem serial number, date of purchase: ____________

____

Modem warranty length/terms: ___________________

Modem manufacturer's phone number: _____________

Do you have a router? Y/N: _____________________

Router manufacturer/model number: ______________

____

Router serial number, date of purchase:

____

____

Router warranty length/terms: ______________________

Router manufacturer's phone number:_______________

If wireless, network name and SSID: _________________

____

Does the router include a firewall? Y/N: ______________

If firewall was purchased separately, firewall manufacturer/model number: _______________________

____

Firewall serial number, date of purchase: _____________

____

Firewall warranty length/terms:______________________

Firewall manufacturer's phone number:______________

Did you install antivirus software? Y/N: ______________

Antivirus software key number or account number, if given: ____

____

Installed auto-updates for anti-virus software? Y/N:_____

____

Type of backup:____________________________________

____

Backup schedule: ______________________________

____

Installed auto-updates for general computer software, i.e. Windows? Y/N: ______________________________

____

Installed anti-spyware? Y/N: ______________________

____

Name of anti-spyware product: ____________________

____

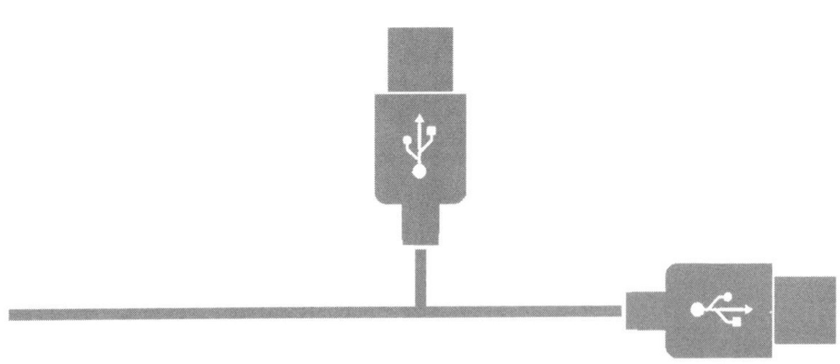

3 Telephones: Landline and Cellular

According to a research study by the Pew Internet & American Life Project, 7 to 9 percent of Americans use a **cellular** or **cell phone** only, with no so-called **landline phone**. Most Americans have some combination of cell and landline service.

A cellular phone is part of a network of radio transmitters and receivers. It broadcasts and receives small-power digital radio signals between the phone and **cell sites** or **towers** located throughout populated areas. Older phones transmitted analog-only signals, and some worked on both analog and digital networks. As of early 2008, however, cell providers no longer had to support analog networks, and all nationwide providers stopped offering this service.

Cell towers are familiar fixtures on the landscapes, but in rural areas, coverage is less available as the networks may not have adequate cell towers.

A landline phone is a phone connected via telephone lines to a central local utility. Local regulatory commissions usually set the pricing of landline phone service. Also, phone service providers are no longer limited to the traditional phone companies.

Your local cable TV provider might also provide both local and long-distance telephone service. The phone service from cable companies is usually a digital signal available over their cable service lines.

You need to understand that phone service from your cable provider is dependent on your cable service being operable and having electrical power. Most cable companies install a battery to handle short power interruptions, but without power, the phone service will not function. Traditional phone service from your local phone company is powered separately from the power transmission lines and so usually works even with extended power outages. If you choose a cable provided phone line, you need to be aware of this major difference.

When considering whether to go with cellular or landline phones (or both), take a close look at the pricing advantages and disadvantages and service and safety issues.

Voice over Internet Protocol (VoIP)

VoIP (Voice over Internet Protocol) transmits your voice calls over your broadband Internet connection. It relies entirely on the quality of your Internet connection, so it is most often used as an add-on service instead of a replacement for your landline. Some additional equipment is required, such as a specialized phone handset. Even the most reliable of the VoIP providers recommend that you keep a backup landline for emergencies and for use with home-alarm systems. VoIP's main advantages are low prices on long-distance (especially international calling) and the portability of your phone service and features.

Pricing

Landline Phone Service

Landline phone service is usually structured to have either "limited calling plans" or "unlimited calling plans" for all local calls. There may be add-on services, each with a fee, for call-waiting, three-way calling, voice mail, and other features. Generally, most homes have a standard phone plan with unlimited local and toll-free calls. If you use a phone for mostly local calls, you do not need to pay attention to the number of minutes your household uses.

Long-distance plans are also available from your local phone company or any number of **long-distance providers**. Pricing on long-distance plans depends on a variety of factors such as whether or not you dial internationally and at what time of day. Phone companies offer a wide variety of long-distance plans, or you may choose to use prepaid calling cards to make affordable long-distance calls on a pay-as-you-go basis. With most traditional phone providers, there are no contract obligations for length of service (as opposed to most cell phone plans that have one- or two-year contract terms). If you have landline service, you can also control your spending on long distance by buying long-distance calling cards from retailers. These are cards where you access the card company's network via a toll-free number, and obtain the rate indicated on the card for long-distance calls.

Cell Phone Service

Cell phone plans are sold on the basis of "minutes used." There are two types of cell phone plans: prepaid or pay-as-you-go plans and monthly service plans with contracts requiring a set number of months or years of service. Most services, such as call-waiting and voice mail, are included in the pricing, as are long-distance charges. Additional features such as **texting** (also known as **SMS**), **mobile web-browsing**, music and video downloads, customized **ringtones**, and sending and receiving pictures on your phone all incur additional charges.

Plans can be customized to cover almost any expected needs or projected usage, but if you exceed the number of minutes or use additional services, the á la carte fees for those additional minutes or services can quickly add up. Toll-free calls still require the spending of your minutes. Calls are usually free at night (the definition of nighttime has start times ranging from 7 p.m. to 10 p.m., depending on the provider) and on weekends. International calls are charged additional fees in addition to the minutes used, unless you have an international calling plan. Also, unlike landline plans, minutes are usually charged on both incoming and outgoing calls (some cell phone plans are the exception to this rule).

Many consumers use prepaid phones as a way to avoid contracts, control spending on their cell service, or obtain service without credit checks. Typically, you purchase a phone with a set number of minutes. When you activate the phone it has all of the usual features of other cell phones. Instead of paying a monthly fee and signing a contract, when you spend your allotment of minutes, you recharge the phone by purchasing additional minutes over the phone with a credit card or by purchasing cards for additional minutes from your cell phone carrier at retailers such as convenience or electronics stores.

Availability of Service

Landline Phones

One of the achievements of the last century has been making phone service available to almost every home in the country. When you consider that this requires connecting a physical line to each home, this is an amazing accomplishment. Phone lines have small electrical currents delivered through their lines enabling them to work even when the power goes out. A reliable signal is, therefore, almost always guaranteed when using a landline phone.

Cell Phones

Cell phone service allows you to move beyond the limits of traditional phone lines. However, there are some considerations to keep in mind when evaluating cell phone service.

You will only be able to make and receive calls within a designated **coverage area**. Cell sites or cell towers transmit signals within the coverage area. If you're close to one of these sites, you can easily receive and make phone calls. (Don't worry—you don't need to always know where these sites are or if you're close to one. But if you're not getting a signal, that could be the problem.)

The point is—it's important to choose a provider that has the best phone network coverage in the area where you plan to use your phone. A cell phone provider that has great phone selection, affordable pricing, and friendly customer service is useless if you cannot make a call due to tower-distance problems. Each of the providers features a **coverage map** on their website. Here are links for the top four providers:

- **Verizon Wireless:**
  http://www.verizonwireless.com/b2c/CoverageLocatorController
- **AT&T**
  http://www.wireless.att.com/coverageviewer/
- **Sprint**
  http://coverage.sprintpcs.com/IMPACT.jsp
- **T-Mobile**
  http://www.t-mobile.com/coverage/

These maps provide general guidance. All of the providers have so-called **dead zones** where network service is unavailable even though it's listed as being within their coverage area. Users can report these dead zones at various websites (for example, www.deadcellzones.com).

Other issues can affect signal reception such as the construction of the building itself (thick walls, materials used, other interfering radio waves) and the physical geography of the area from which you are calling. Generally the higher you are, the better the access to coverage. If you are in a valley or at the bottom of a hill or mountain, the radio signal may be obstructed.

In all cases, do your research before choosing a provider by speaking with your neighbors, co-workers, and friends. They will be able to describe their experience with a particular provider, and you can see if one will be best for the places you are likely to be. Many providers will allow you to test-drive their service. Ask if you have a time period to see if the phone network works where you drive, work, live, and spend your leisure time.

Cellular Networks in the U.S. and Overseas

Worldwide, there are two major types of networks: the **CDMA** family of networks and the **GSM** family of networks. CDMA is the name of the wireless network used by most cellular networks in the U.S. and some other countries. Most of the rest of the world and two of the major providers in the U.S. use the GSM technologies (AT&T and T-Mobile). If you only plan to use your phone in the U.S., you can choose the right cell provider for you based on coverage, rate plans, and phones available, without worrying about whether it is a CDMA phone or a GSM phone.

If you wish to use your phone overseas, you may want to consider a GSM phone. (By the way, cell phones are usually called "mobile phones" overseas.) But even within GSM countries, there are different frequencies, so be sure to ask your provider about usability in the particular countries that you are visiting.

Generally speaking, GSM phones don't work on CDMA networks and vice versa. Many phone manufacturers and cell providers are finally solving this dilemma for their customers who travel internationally by offering both types. These are called **multi-band phones** or **global phones**.

Emergencies

In most power outages, due to weather or system failures, your landline telephone will probably still work. Each phone line to a home carries a small electrical charge that powers the system independently of your power company. Of course, your cordless phones *will not work* as the base requires electricity. Every home, therefore, should have a **corded** or **analog phone** available for use in such a situation.This corded phone doesn't have to be connected all the time—you can store it with your emergency supplies and exchange it with the cordless phone when necessary.

If you have landline phone service from a non-phone provider (such as cable), you should be sure to ask about their plans for power outages and emergencies. Most major cable providers have power backup generators and Enhanced911 or **e911** service. E911 allows the emergency services to give a physical location of a phone caller by using a reverse lookup so that even if the caller cannot provide their address, the authorities at the 911 call center can know where to respond.

Only traditional landline phone service will have power in most situations when the power is out, as the phone line themselves carry a small electrical charge. Phone service from your cable company will most likely have a battery backup that should last for a few hours when the power is out. It is important to know how long or even if your phone will work in a power outage.

In major emergencies (such as a hurricane, flood, or tornado), the landlines might also lose power or be damaged. In these very severe situations, it may be necessary to evacuate or leave your home. In these instances, your cell phone might be an essential tool in maintaining communication. Remember, however, that in large-scale power outages, your cell phone may not work as the cell phone towers may be without power or damaged themselves. In all cases, you should have a **car cell phone charger**, as you will not be able to recharge your cell phone with your standard charger if there is a widespread electrical outage.

All modern cell phones must also be e911 enabled. Unlike with landline phones, in which a reverse lookup is used by emergency operators, a cell phone caller can be located either by **GPS (Global Positioning System)** chips in the phone or by detecting the phone's location by looking at which cell tower or towers the phone is closest to. In some areas of the country, however, the 911 call centers may not have the equipment to use the e911 information. Any cell phone that can receive a signal, even those without an active service plan, should be able to dial 911. This is why it might be a good idea to keep a charged phone in your car glove compartment in case of an emergency. Do check it from time to time to see if it turns on and detects a signal.

When dialing a cell phone, even one with e911 capabilities, do be aware that you *may not be connected to the closest emergency provider*. This is especially true if you are calling from an area on the border of two communities. You will be transferred to the correct call center eventually, but it's important to keep this potential delay in mind for those emergencies in which every minute or second counts. You should enter as a speed dial your local emergency contacts, so if you have a local emergency you will reach the correct call center.

In the event that you are traveling and are not able to get a signal, try some of these techniques: try climbing to the highest point of the surrounding area, hold the phone away from your body and over your head, and try using a **text message** (see next section), as this shorter communication may be able to get through when a phone call may not.

One last note about emergencies: it is a good idea to enter a contact phone number into your cell phone's address book under the heading of **ICE (In Case of Emergency)**. After the word "ICE," enter the phone number of the individual you have designated as your emergency contact. That way, if you are in an emergency and are unable to speak for yourself, EMT (emergency medical technicians) or police officers on the scene will know whom to contact.

Cell Phone Features

Text Messaging

Text messaging (also known as **SMS** or **Short Messaging Service**) is a method of transmitting short messages, usually up to 160 characters, over cellular phone service. Text messaging can be a very efficient way to communicate and can be just the thing to get a short piece of information to your friends, family, or co-workers. Texting is quickly replacing e-mailing as it doesn't require logging into your computer to check e-mail.

A whole new vocabulary called "**textalk**" has developed around texting. As you are limited in message length and as typing on the phones is more difficult for many than typing at a keyboard, all sorts of shortcut words and phrases have become standard. One good resource listing all of the most common terms is available from NetLingo (www.netlingo.com).

SMS alerts are becoming the primary method of transmitting emergency information among communities such as a university or large companies. These short messages take up less of the cell networks capacity and so are more efficient than multiple phone calls going out over the same network. In many mass emergency situations, SMS messages were able to get through despite overloaded cell phone networks. The **FCC (Federal Communications Commission)** just approved the development of a nationwide alert system. A federal agency will create messages that will be broadcast through cell phone providers in the event of a national emergency such as an imminent terrorist threat, weather emergency, or child abduction alerts. The plan is targeted to be in effect by 2010.

Text messaging is usually an "added feature" to a cell phone plan. If you haven't purchased a set number of messages, texting can quickly become expensive with an average charge of 20 cents per message, both sent and received. Be aware of the cost of premium text services such as picture messaging. These are messages with embedded pictures, music, or video, and these incur even bigger charges.

Internet Browsing

Most new phones, even the most inexpensive ones, will have the ability to browse the Internet. Many of the most popular sites have customized their pages to be viewable on the much smaller screens of a phone versus a full-sized monitor. Depending on the type of phone you have, you will be using one of the following technologies:

EvDo (for CDMA phones), 3G (for some GSM users), EDGE (for other GSM users), or Wifi (for any wifi-enabled smart phone).

Wifi is only available for local area surfing and is generally free to connect to and use with the proper permissions to access an existing network in one particular location (like a library, airport, hotel, or restaurant). The other technologies connect to the Internet without requiring a separate Internet connection. As you are using your cell provider to connect to the Internet, you are required to either subscribe or pay a per-use fee. These fees can be quite expensive and are usually charged by the kilobyte. Simply watching an Internet video could result in exorbitant charges on your next bill.

You should check with your cell provider about pricing and availability of Internet browsing on your phone. Usually it is listed under a **data plan**. A data plan refers to coverage for receiving and sending web-based e-mail and Internet browsing. This is an expensive service to add (from $20-$100 a month) but even more expensive to use if you do not have a plan that covers these services, where they will charge you by the kilobyte. If you are considering doing any sort of web browsing on your phone, make sure you fully understand this when buying a cell phone and plan.

E-mailing

Many providers and cell phones will allow you to check and send e-mail from your phone. This is separate from your SMS (texting) capability. This is usually part of the data plan, but check with your provider on options and pricing.

Customized Ringtones, Music, and More

Your cell phone provider has an array of downloadable content for your phone. These can be songs, TV shows, video clips, games, ringtones, and other media. Most of these are sold as individual items that are downloaded to your device and are usable on that device. Some providers have multimedia subscriptions that allow for listening to a wide variety of music and videos without saving to your phone.

Managing Your Cell Phone Bill

Choose a plan that suits your needs. Before you start a contract, be sure you are aware of what is in your particular plan and what isn't. All of the major providers will hold you to an "early termination fee" that may be hard to avoid if you are unhappy with your service. You should know the *length of your contract*, and if you are extending your contract by purchasing a discounted phone or changing plans. We suggest making this an explicit question before signing or agreeing to anything.

Only choose a plan that allows you to *test-drive their network*, be it for two weeks or thirty days. You should be able to cancel a contract if the service does not meet your needs.

As of late 2007, most of the major providers will allow you to upgrade or downgrade your plan and its features without extending your contract. This is an excellent opportunity for you to analyze your monthly bill to see if you should be adding or dropping minutes, and adding text messaging or a data plan. If you consistently don't use the minutes in your plan, and your minutes needed would fit under the next lowest pricing plan, you can save significant money over the life of the contract. One piece of advice is to check the provider's online account page to see how many minutes you have used if you find yourself relying on your cell phone more than usual. Often, you can upgrade your plan before the billing cycle ends and avoid paying overage fees. This higher-than-usual use of your cell phone can happen when a relative gets ill, during other family emergencies, or when planning events such as a vacation.

Before you travel, be sure to find out if you will be paying **roaming**, out-of-network, or international charges by using your cell phone, text messaging, browsing the Internet, or e-mailing. A quick call could save you the shock of a bill in the thousands of dollars at the end of the month.

Landline Phone Handsets

Landline phones come in two main types: **corded** and **cordless**. As stated above, every home should have a corded phone on hand in the event of an emergency. Corded phones are analog. Analog phones tend to have better sound quality, but they are susceptible to static and interference from other electronic devices.

Cordless analog phones are *less secure* than digital phones as they can be eavesdropped on. Cordless analog phones usually can only support two handsets in a home; digital phones can support multiple handsets. We recommend that new phone purchasers look at **digital phones** for their security and for available features.To be sure you're actually getting a digital model, check the packaging carefully. It should say "digital phone," "DSS," "frequency-hopping spread spectrum (FHSS)," or "digital enhanced cordless telecommunication or telephone (DECT)."

You may have heard that cordless phones can interfere with home Wifi networks, baby monitors, and other electronic devices in your home. Earlier cordless phone systems used various radio frequencies to transmit phone signals throughout the home.The good news is that this problem has been addressed by the Federal Communications Commission (FCC) by reserving the 1.9GHz frequency for voice products.These phones hold a battery charge longer, allowing for longer standby and longer talk times, and can support up to ten handsets.

When shopping for cordless phones, you may still see older models that operate at 900MHz, 2GHz, and 5.8GHz at low prices, but it is best to shop for the newer 1.9GHz DECT phones. As these phones are now the prevailing standard, their prices should begin to be more competitive.

Different features that you may want to consider are:

- the ability to add or replace additional handsets to your base system
- a phone answering device
- and the ability to conference multiple handsets on a single phone.

Some phones also offer the ability to:

- connect to a Voice over Internet Protocol **(VoIP)** connection
- connect your cell phone to the landline phone system using wireless **Bluetooth** technology to answer both phones interchangeably or to use whichever phone offers the best use of your plan minutes or long distance.

Cell Phone Handsets

Cell phones have evolved quickly since they first arrived on the scene in the mid-1980s as expensive new business technology. A cell phone can be anything from a large clunky phone with a monochromatic display to a full-featured miniature PC with web browsing, video streaming, and video recording. When shopping for a cell phone handset it is useful to think of them falling into three general categories:

- Basic cell phones—Even the most introductory models are feature laden. Most include a camera, options for customizing backgrounds, and different preloaded ringtone selections. Providers even make games and customized ringtones available for purchase.
- Media phones—These are hybrid devices that function as both a basic cell phone, but add the ability to store and play music and videos and perform some Internet browsing. Some of these phones may also have QWERTY keyboards (a key setup that resembles a computer keyboard's) for easier text messaging.
- Smart phones—These combine the functions of a

phone, personal media player, and PDA (personal digital assistant).The most modern of these have come to be tiny portable computers with e-mail, office software suites, and full Internet browsing functions.

When deciding which cell phone to choose, consider:

- what you want your phone to do
- which available phones work with your provider
- and how much you want to pay.

If you're on any kind of a budget, having a cutting-edge "smart phone" with a full keyboard and an MP3 player is going to make living within your means very difficult with all the temptations to surf the web, text, and download songs.Your children might want you to indulge them with the latest model. But doing so may cost you a fortune when they send videos to their ten closest friends—and you get the bill.

Read reviews of phones you are interested in buying, especially if none of your friends or co-workers have that particular brand. Good sites for reviews include:

- C-Net.com
- Engadget.com
- Gizmodo.com

Features to Consider

Type of network: Is it a GSM or CDMA or a phone that can switch between both networks when traveling? See page 47 of this chapter for more information.

Size and shape of the phone: The phone should be sturdy and easy for you to use. Check to see how the phone feels in your hand. Does it cradle comfortably against your ear? Can you see the menus? Are the buttons easy to read, and are you able to press them independently? Do you want a flip phone to avoid accidental misdials?

Does the phone support Bluetooth? Does it accept corded headsets? Have you seen people who appear to be talking to themselves, and then you notice a funny-looking gadget stuck to their ear? That's a Bluetooth headset. Bluetooth is a method of transmission between two devices that are "paired." It uses wireless technology like Wi-Fi to transmit information between objects that are Bluetooth-enabled and located in close proximity to each other. One of the most common uses of Bluetooth is to provide wireless headsets for your phone, allowing you to have hands-free use of your phone. Other users may prefer a corded headset.

Does the cell phone you're considering give you the option of using either Bluetooth technology or a corded

Geekspeak: Bluetooth

Bluetooth is a low-power, close-proximity wireless protocol that creates a Personal Area Network (PAN). While most users are familiar with this technology for creating wireless phone headsets, there are many other uses: headphones for listening to your personal media player or for your computer audio; data transfer between devices like your computer and printer or from your computer to a portable storage device; wireless game controllers and computer peripherals (like your mouse or keyboard); and for linking your cell phone or GPS into your car's dashboard controls.

headset? With both types of headsets, does the phone support **voice dialing** to go completely hands-free?

Is the phone able to perform "push to talk" functions? "Push to talk" acts like an intercom over the wireless network. If so, does the plan cover this option?

Web browser: Be sure to see what type of browser comes pre-loaded and if it is one that you are satisfied with. If you are considering a phone with a web browsing feature, it is an expensive purchase so take the time to do some real browsing on the device to see if it is easy for you to use. Ask if you can download a browser that you are more comfortable with such as Internet Explorer, Firefox, or Opera.

E-mail: Can the phone check your e-mail and send e-mail? Is it compatible with your work's e-mail system? It is a good idea to check with your company's IT department if you want to use your phone for e-mail.

Text features: Does the phone have a standard QWERTY keyboard, similar to your home computer keyboard? Are the keys easy to use for both symbols and letters?

Media storage: Can it act as a replacement for your portable music player so that you will have one less device to carry? Can you only load music through your provider?

Camera and video camera features: Does your camera have these features? Do you want them? Are these features easy to use? Will you be charged for getting your pictures off your phone?

Other options: Some phones have built-in FM radio receivers, calendars for organizing your schedule, alarms, multiple time settings, and extensive phone books.

Teens and Cell Phone Use

Teens and cell phones open up a variety of concerns. Parents like to be able to keep track of their teens and busy schedules; teens like to keep up with their friends. A few points should be considered to keep your teens safe and your budget intact.

When and where they can use their phones. Schools generally allow high school students to have a cell phone, but place restrictions on cell phone use. Some require that they be turned off during the school day; others will allow students to carry their phones but with their rings silenced. In effect, many teenagers do use their phones in spite of the regulations, mostly for texting their friends. You should make clear to your teen that they need to follow school policy or their phone may be confiscated and kept for parent retrieval from the front office. A discussion about not using the phone for cheating may open your eyes to what your child has seen go on in classrooms.

Cell phones and driving. We all know that driving while on a cell phone can lead to distracted driving, even with headsets. Worse yet, many teens have picked up the habit of texting while driving. Laws are being considered to ban this practice nationwide, but discussions between parents and their children shouldn't wait until laws are passed.

Excessive Texting. Beyond the financial shock of finding that your teen went hundreds of text messages above your plan's limit, excessive texting can be a sign of a troubled relationship. Texting can interfere with their sleep and attention in school, and might be a sign that one of their relationships is controlling or abusive. Predators realize that parents are less likely to be aware of the content of text messages, especially as they can be done discretely.

Financial Peer Pressure. Many teens have unrealistic expectations of the reality of budgets and how much cell phones and their features cost. They are likely to want the latest phone, with full features such as Internet browsing and video recording/playing. Often, their friends will send out texts not realizing the your family's plan is limited. Or they send pictures and attachments that can incur charges at the receiving end. Many parents are shocked to find out that receiving a message like this even if you don't open or look at it still incurs a charge.

Premium cell phone services from third-party vendors. If you watch any show aimed at a pre-teen or teen audience, you will see ads for ringtone providers, joke services, and all sorts of subscription services that hide their true cost in the fine print. With no adult consent, your child could unwittingly add a "service" that sends out worthless text messages with a monthly subscription cost. Your cell phone provider just passes these charges along to you and requires that you talk directly with the third-party vendor to prevent these charges from coming back month after month.

You have the ability to block text messaging, premium text messaging, and other services from handsets on your bill. This is usually done at no charge and with no negative effect on your contract. If the temptation for your teen to abuse your trust is too great, a pre-paid cell phone might be a good idea, or even, banishing the cell phone entirely.

What to Do with Your Old Phone

As with all electronics, there are multiple hazardous chemicals in your phone and its battery. You have two main options in recycling your old phone: donate or recycle. Many charities accept phones and then either refurbish them and make them available for reuse or decide that the phone is outdated and recover the reusable components. You can contact your cell phone provider for any programs that they have in place, research charities on the Internet that collect old cell phones (www.charitynavigator.org will help you determine if the charity is an effective one), or contact your local government to see what their electronic recycling program is. In any case, it is a good idea to clear your personal data off the phone (see www.recellular.com/recycling/data_eraser/default.asp for information on how to clean your particular phone).

Checklists

Landline Phone Service Comparison Chart

Provider Name			
E911 Y/N?			
Works when there is a power failure? Yes, but with battery. Or, No.			
If you have a home alarm, is it compatible? Y/N			
Monthly rate for unlimited local and toll free. Is there a low-use plan (limited calling)?			
Call waiting/ call forwarding/ voice mail			
If battery, battery life?			
Other:			

Landline Phone(s) Records

Your phone number:______________________________

Type of service (traditional/digital): _________________

Will it work when power is out?: ___________________

Corded phone for emergencies? Y/N

Cordless phone system? Y/N

Manufacturer/Model number: ______________________

Serial number and date of purchase:_________________

____

Warranty period?_________________________________

Operates on what frequency? (see manual) ___________

Cell Phone Carrier Comparison Chart

Provider Name			
GSM or CDMA network?			
Coverage map okay for near home/work/travel areas?			
Plan minutes?			
Contract length?			
Free calls?			
Free nights/weekends and start time?			
Family plans; how much for additional phone?			
Text messaging per message charge?			
Unlimited text messaging plan cost?			
Internet plan charge?			
Picture message charge?			
Any bundled plans?			
Phone/text/Internet plan?			
Early termination fee? If so, pro-rated?			
Test-drive of system available?			
Other:			

Cell Phone Provider Records

Your chosen provider's name: ______________________

Your cell phone number: __________________________

Customer service number: _________________________

Additional numbers on plan:

____

____

____

Cell Phone Handset Comparison Chart

Provider Name			
Price with contract?			
Price without contract?			
Single band or world phone?			
Bluetooth? Y/N			
Speakerphone? Y/N			
E-mail capable? Y/N			
Internet browser? Y/N			
QWERTY keyboard? Y/N			
Battery life?			
Camera phone picture size? (megapixels)			
Built-in flash for camera?			
Plays music? Y/N			
Media-storage size?			
Other?			

Cell Phone Handset Records

Manufacturer/Model number: ____________________________

____

Serial number and date of purchase: ____________________

____

Warranty term: ____

____

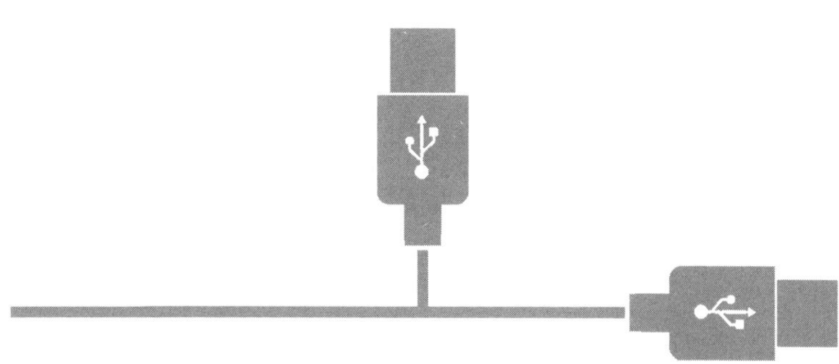

4 Television

For years, most changes in television technology related to the hardware side of things—larger screens, High Definition (High-Def), and advanced features. Today's big changes involve how we receive our content. Now satellite and fiber optic transmissions, video streaming, and websurfing on the Internet are all leading to better picture quality and a more interactive entertainment experience.

Analog Is Dead—Are You Prepared?

All analog TV broadcasts will stop on February 17, 2009. After that date, any device that relies on an over-the-air signal (television or radio that receives the sound portion of TV broadcasts) will not work without some sort of digital-to-analog converter box.

Why? According to the government, "Broadcasters are transitioning to digital to provide important benefits to consumers. Because digital broadcasting is more efficient, broadcasters require less of the airwaves to provide a better television viewing experience. Once the **DTV transition** is completed, some television channels will be turned over to fire and police departments for emergency communication, and others will be auctioned to companies to provide new wireless services." (https://www.dtv2009.gov/)

Now that's quite a statement—but what does it mean? First, let's assume that you have an older TV and that it is not digital ready. In that case, you have a TV that can only receive analog TV.

Analog TV

Analog TV is the standard TV broadcasting that you can receive with a **roof-top antenna** or a set of **rabbit ears**. This signal will stop on that date. Does this mean that everyone's TV sets will have no signal? No, this will affect those who just have over-the-air antennas only.

For those customers who have cable TV plugged right into your TV—you will not have any problems. If you have a cable TV converter box, you will have no problems. If you have satellite TV, you will have no problems. It only affects you if you use an antenna to receive your TV signal or radios that receive the sound portion of TV broadcasts.

Early users of the over-the-air digital broadcasts have found that the digital transmissions are of much improved quality over their old analog TV broadcasts. They have also found that there is more programming available compared with the old analog TV lineup.

In addition, many TV viewers who rely on basic cable for local content may find that they can do without cable after converting to digital TV. In most television markets, you do not have to wait for the February 2009 date to try receiving these digital broadcasts.

Adding a Converter Box

If you want to keep using your current TV and antenna, you do have an option. You will need to get a converter box. The box will convert the new digital signal to a signal your old TV can use. The price for these boxes is around $60, but as part of the transition to digital broadcasting, the U.S. government has a coupon program to offset the cost. The coupons are good for $40 off the price of a converter box, and each household can receive two coupons. Visit https://www.dtv2009.gov/ to order, or call 1-888-388-2009.

If you use **closed captioning**, the analog-TV-plus-converter-box setup should still work for you under most circumstances. The only exception is if your TV was built before 1993 or if the screen is smaller than 13 inches. If those exceptions apply to you, visit http://www.fcc.gov/cgb/consumerfacts/CC_converters.html to find instructions on how to keep receiving the captioning.

Battery-Operated TVs and Emergency Situations

One major thing to keep in mind if you have a battery-powered TV for use in an emergency—you will not be able to use the converter box on this type of TV as the converter box needs electrical power. As of yet, there are no battery-powered converter boxes. You will need to buy a battery-powered *digital* TV, which may be difficult to find.

Digital Ready

How do you know if you have a TV that is digital ready? By law, all TVs made and sold after March 1, 2007, must be digital ready. For older TVs, you can check your owner's manual or look on the TV itself. All digital ready TVs will have an input labeled either "digital input" or "**ATSC,**" which stands for Advanced Television Systems Committee.

If you bought a TV after 2004, you have a good chance that your TV is digital ready. But not all TVs sold after that time are digital ready. Some TVs are simply display monitors (like a computer monitor) and don't have the ability to decode either analog or digital TV signals; they rely on a cable or other converter connection. These TVs were described as "HD-ready" or "HD Monitor."

High-Definition or High-Def TV

Warning: This is a potentially TDDZ (Too Digitally Daunting Zone), proceed with caution and with full permission to nod and say, "yes, yes, very interesting," before moving along. It may be a bit too much information, but it's also helpful in understanding the basics of why a certain HDTV is better than another. If it's too much . . . skip ahead. We'll see you after all the acronyms and numbers!

* * *

Before we explain **HDTV**, it helps to understand what "standard definition" or **Standard TV** is and have a quick understanding of how a television screen works.

What Is Standard TV?

Standard TV has 525 horizontal lines of picture (a line of **pixels**—points of light), of which 480 are visible. The picture on a standard TV set is redrawn in a method called interlacing. Interlacing means that each time the picture is refreshed, it redraws the picture on every other line, thereby lacing together one image with the next. It happens quickly enough that your eyes don't notice the process. Standard TV is also called a 480i (480 lines interlaced) TV.

What Is Enhanced TV?

When the concept of higher definition TV was first introduced, the first step was something called Enhanced TV. This standard has the same number of pixels as standard 480i TV, but the rows of pixels are not interlaced. They are instead "**progressive scanned**"—which means each image is completely redrawn on the screen. This is referred to as 480p (the "p" stands for progressive).

What Is Hi-Def or HDTV?

In order for it to be a true HDTV, it must display at least in 720p mode; this means 720 horizontal lines progressive scanned. The next step up is 1080i/p.

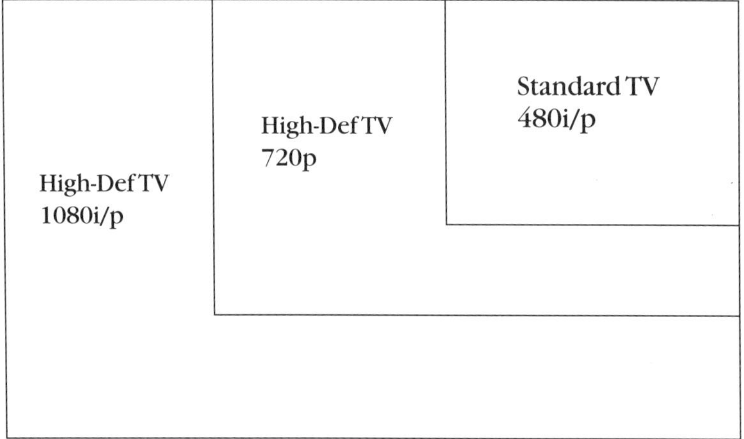

1080p Is Better than 1080i; But Should I Pay Top Dollar for It?

If "progressive scanned" produces a better picture than an interlaced picture, doesn't that mean I should always buy the best, meaning the more expensive 1080p TV? Not always. There are two issues:

- Broadcasts - Just because an HDTV can display a picture in 1080p doesn't mean that the signal you receive from your content provider is of that quality. The broadcasting station may be sending a 720p signal. An example for the sports fan: ESPN-HD broadcasts at 720p because they think it provides a better image for displaying high action. Another factor is that your HDTV signal from the cable company or satellite company may not be at full 1080p.
- Viewing experience - The second factor is whether or not your eye will even be able to tell the difference. Perhaps if you are sitting very close to the screen, you may be able to tell the difference, but remember these are sometimes sixty-inch screens designed to be viewed ten to fifteen feet away or more.

Rear-Projection CRT (Cathode Ray Tube)

A **rear-projection CRT** uses the same technology that your old TV used but updated to produce an HD picture. The great thing about this type of set is that it is proven and reliable. The main drawbacks are: they are extremely heavy, and they are not a big step forward. They have a bulb that will at some time need to be replaced, but you can expect the bulb to last around four years if you watch four hours of TV a day.

Another caution is that these screens can suffer from "screen burn in." That is when an image is displayed without moving for some extended period of time, and it "burns in" to that part of your screen. This might happen with the little network icon that never leaves the broadcast image on some channels. It is a rare phenomenon, but something to be aware of. Also, depending on the model, the viewing angle may be smaller than with more expensive TVs. This means that the picture loses quality the farther away the viewer is—right or left of the center. The major attraction of these sets is the price; a good 52-inch set is less than $1000.

Pros:

- Inexpensive

Cons:

- Big and heavy
- Bulb replacement required eventually
- Viewing angle tends to be smaller
- Susceptible to burn-in
- Warm-up times of twenty to thirty seconds

Digital Rear-Projection TVs

Rear-projection LCD, DLP, and LCoS are a step beyond the CRT. These TVs reflect light off a LCD (Liquid Crystal Display), or through a DLP (Digital Light Processor), or via a LCoS (Liquid Crystal on Silicon) chip onto a big screen. As with the CRT TVs, bulbs may need to be replaced after a few years of viewing. The advantages over CRT are that these TVs look better in brighter light conditions than CRT, and they are much lighter and less bulky. They also can display digital signals, unlike CRT TVs.

Manufacturers are investing heavily in this segment of the market, so expect to see improvements, especially in the light source.

Pros:

- Very low price for a large screen
- Can process digital signals
- Manufacturers are investing heavily in them

Cons:

- Currently, DLP TVs cannot display 1080p
- Bulb replacement required eventually
- Small viewing angle (about 128 degrees)
- Warm-up times of twenty to thirty seconds

LCD TVs

LCD stands for "Liquid Crystal Display." Instead of projecting light off an LCD chip, these TVs are composed of a full LCD display. One of the many advantages to LCD TV is that there is no flicker as there is with a CRT or standard TV. Another advantage is that they have a matte finish, which means there is little glare in a room with daylight. This is much easier on your eyes. Unlike standard TVs, you can see a great picture no matter where you are in the room—they have a 160-degree viewing angle.

While the picture is not as good as a plasma set and the screen sizes are not as big, LCD TVs are quite good and getting better. They are also usually cheaper than plasma sets and use a lot less power.

Pros:

- Good color reproduction and improving contrast
- Very thin, and getting thinner
- Relatively lightweight, with flexible mounting options
- Excellent longevity
- Among the brightest direct view displays
- No practical screen burn-in effect

Cons:

- Very difficult to produce deep blacks (LED backlighting is improving this, however)

Plasma TVs

The technology involved here is quite difficult to explain. (But if you must know, the picture is created by excited cells containing noble gases excited by electrons to produce different colors of light). What you need to know—the result is a high-quality display that is bright and has a great picture.

Pros:

- Excellent life expectancy
- Excellent color reproduction, including black levels
- Excellent viewing angle with no real loss of color or contrast

Cons:

- Fairly heavy
- Thicker than LCDs by a large margin
- Susceptible to screen burn-in
- Uses a lot of power compared to an LCD

Choosing the Right TV for You

Now you know a little about what types of TVs are available and the pros and cons of each type, but how do you know if a particular TV will work in your house? What options do you *need* and what options would be nice to have?

Here are some things to consider:

Room Size

The size of the room will help you decide just how big a set you will want to get.

The maximum screen sizes on this chart can get to be a little big and can only be achieved with a projector. I will cover these later, but you get the idea—you don't want a 52-inch TV screen when you are going to be four feet from it.

Viewing distance from screen in feet	Min Screen Size in inches	Max Screen Size in inches
4	19	32
6	29	46
8	32	63
10	40	80
12	46	96
14	52	112

Seating Placement

Most TVs have an optimum viewing angle (the picture will look better for someone sitting right in front than for someone sitting at an angle to the set). Placing the TV where everyone has a good seat is always a good idea.

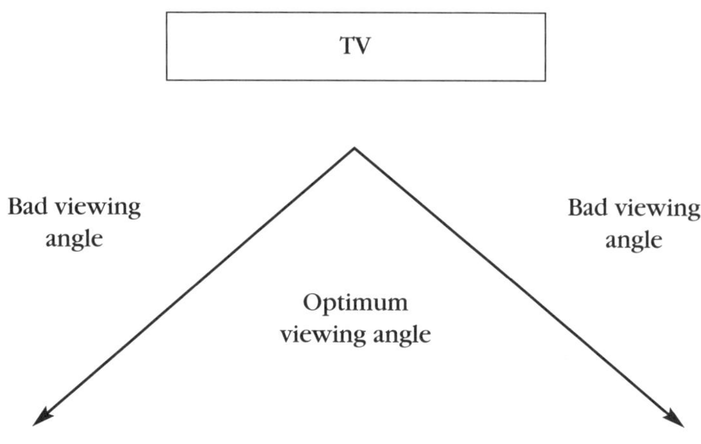

Good and Bad Viewing Angles

Sound Quality

The speakers that come with most TVs provide pretty good sound, and some even attempt to provide a surround-sound experience. While you don't *need* to get a better sound system, here are some things to keep in mind when making decisions:

Surround Sound

This is when you surround the viewer with sound via speaker placement and special equipment used to produce a theater-like sound experience. The most common type is called "5.1 channel surround." This involves the use of five speakers and an amplifier that supports it.

The TV would be placed under the center channel speaker. The rear speakers are usually much smaller than the front ones. If you watch a lot of movies, this will add greatly to the experience. However, this is an expensive option and something you can add later if you wish. It's not necessary to have this to enjoy movies and sports.

Surround-Sound Arrangement

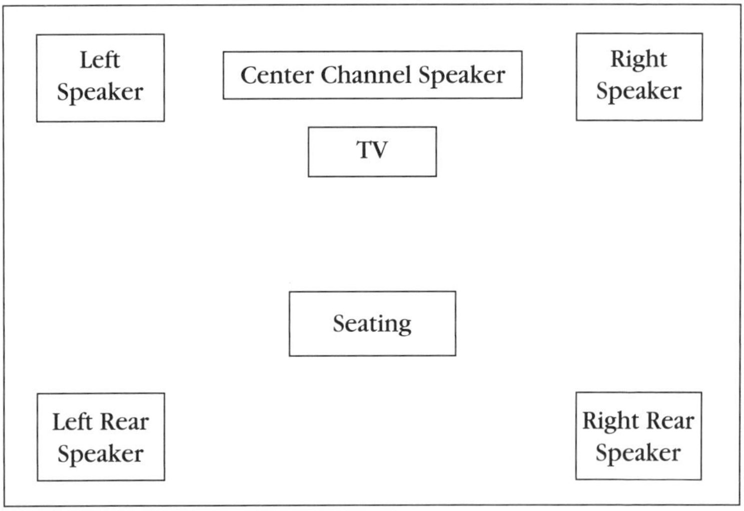

What Are All Those Plugs and Cables?

Hooking up a TV can be a daunting process. First, if you are buying a TV that you want to mount on a wall, you really should pay for professional installation. If not, you risk doing damage to the walls of your house and to the expensive new TV you bought. There are issues of placement as well.

You should not mount a TV over a fireplace as the heat can damage the electronics or mount it anywhere there is the possibility of water damage. You want to make sure there is power and a TV signal in a nearby location.

If you are buying a TV that will be in an entertainment unit or that has its own base, once you get it in the door you are halfway there. If you are Digitally Daunted, you also may be intimidated by the back of any TV—with all of those receptacles and color coding. Open the manual and let's get to work. If you take your time, look at everything carefully, and label all the cables, everything will be okay. Here's the information you need to know to get started with confidence.

Shapes and Colors

Most cables are color coded—you match those color-to-color to hook them up. If they are all black, they will have a distinct shape that will only connect to one place.

Labeling your cables can save a lot of headaches. On each end, wrap some masking tape and with a permanent marker write on the tape what the cable is for: DVD, cable box, stereo, and so on. This way you can hook up one end and feed all the cables to where they are going and not worry about getting them mixed up. It is also helpful if you need to move anything or replace something in your system—you won't need to trace each cable back to its base.

Input and Output Connectors

There are a number of different input and output connectors. Let's go over the different types.

HMDI: High-Definition Multimedia Interface

This type of connection is used for connecting just about everything from DVD players to video game systems. You will almost always need to buy the cable for this connection separately. You will most likely get the hard sell from many stores to buy an expensive cable (up to $150). Don't believe any of the sales pressure on this; there is little difference other than price between the expensive cables you find in the stores and the cables you can find online at places like Amazon.com for about $10. These are digital signals traveling short distances, so the money spent on expensive cables is money that you do not need to waste.

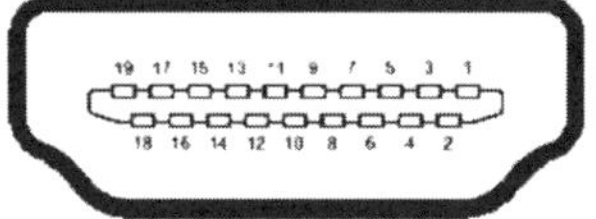

DVI: Digital Visual Interface

This connection is used for PCs and laptops for the video signal only, unlike the HDMI described earlier, which has both the video and audio signals on the one cable. You can also get a converter for one end of the cable so it will fit into an HDMI plug.

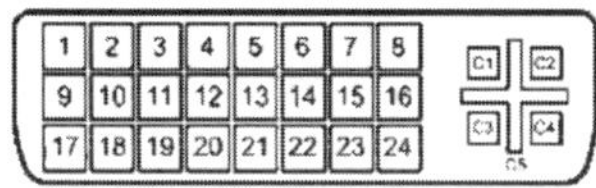

Component Video and Audio Cables

Some TVs have two sets of cables: three cables for the video signal (green, blue, and red) and two audio cables (red and white). All of these cables are RCA-type cables, which look like these shown below.

3 video cables 2 audio cables

S-Video

The next rung down is S-video cable. You will most commonly see this on camcorders and some DVD players.

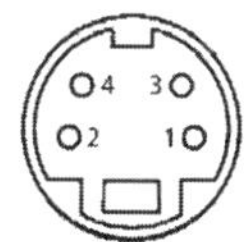

Composite Video

The composite video cable with the yellow, white, and red plugs is the most commonly seen cable. It's used primarily with older video equipment and video games, but almost all video devices support it. The red and white plugs are for audio; the yellow carries the video signal.

(left to right: yellow, white, red)

Getting a Signal

No matter what type of television you choose, you will need a video source so you can watch it.

There are five main ways to get an HDTV signal to your TV.

Over-the-Air Antenna

Believe it or not, the best HDTV signals can be received over the air with an antenna. No, you don't get as many channel choices as with cable or satellite, but you get the best picture and sound by far—and it's free. All cable and satellite providers have to mess around with the signal to get it to work with their systems, so there's always a loss of signal quality. Although, it's minor in most cases.

Pros:

- Free
- Best signal

Cons:

- Requires purchase and set up of a new antenna
- Not as many channel choices

Cable TV

Almost every area has cable TV of some sort, and if they don't have a wide selection of HD content, they soon will. You will have to rent equipment from your cable provider, which adds some complexity to the set up. You will also have to use the cable box remote, which sometimes will not control all of the TV's function, so you will need to keep the TV remote handy as well.

Most HD packages from the cable companies include features such as a **Digital Video Recorder (DVR)**, which allows you to select programs in advance to record from various television station schedules. With a DVR set up—when you are watching live TV—you can skip past commercials, make your own break times, rewind to catch something you missed, or get season passes of your favorite shows. Another option offered by most cable companies is "On Demand" movies, which allows you to watch selected movies at any time convenient to you.

You can also get a "**CableCARD**." Most new TVs have an option for the cable company to put a CableCARD in your TV. There is a slot on the side for it to go in, and this will allow you to use your TV's remote control instead of the one provided by the cable company.

The biggest negative with cable TV is not the programming, or quality of the product, it is the variability in customer service. Some local providers are excellent, and others provide service that is absolutely unacceptable. As a consumer, it is useful to remember that your local government regulates them because they have a local monopoly. If necessary, you can always contact your utility commission to lodge a complaint.

Pros:

- More channel selection
- Most include a DVR and On-Demand movies
- Can be bundled with phone and Internet

Cons:

- Monthly fees can get expensive
- Customer service can go from great to unacceptable

Satellite TV

There are two major providers of Satellite TV in the U.S.: DirecTV and the Dish Network. Once again, you will need specialized equipment—both on the outside of your home (a dish) and a receiver on the inside. There is also the unique requirement that you will need to be able to locate the dish in a place that has access to a clear view of the Southern sky. If you are a renter, you may have to have permission from your landlord to install the dish. Professional installation is usually advised.

There is a wide selection of channels, often more than with cable. The satellite receiver can be upgraded to have DVR and on demand programming available.

Cable TV ads often depict satellite reception as spotty, especially during rainstorms. This is true when there are heavy rainstorms. However, cable is also susceptible to the same atmospheric conditions, and, in fact, they receive their programming via satellite to their main offices.

Customer service issues have also plagued satellite companies, as it seems that most service companies try to reduce their support staffing costs.

Pros:

- Channel selection is often better than cable
- Most services include a DVR and On-Demand movies

Cons:

- Monthly fees can get expensive
- Customer service can range from great to difficult
- No bundled services for phone and Internet are available

Fiber Optic

As **fiber optic** service spreads to more urban and suburban areas, it is becoming a popular choice. As we previously mentioned in the Internet section, services such as Verizon's FiOS are available in a bundled package with your TV programming, Internet, and telephone (both landline and wireless).

Fiber optic providers also allow for a true **media PC** environment. If you pay the additional fee for the media PC, you can network your photos, music, TV programming, and all other media to all the TVs in your home.

As this is a newer technology, multiple appointments may be required for installation. However, service seems to be reliable after the initial installation is established. The customer service should be on par with the service you receive from your phone company.

Pros:

- Channel selection is very broad
- DVR and On-Demand movies are available
- Bundled pricing available on phone and Internet

Cons:

- Monthly fees can get expensive
- Not widely available

Other TV Content Available

Internet TV Programming

In addition to the popular video-sharing sites on the Internet, there are many Internet TV channels. If you have a **media PC**, you can stream these onto your TV. Most of this programming is free, and a lot of it is international.

A reliable and fast Internet connection is required, and you should be careful about the sites you visit, as with any Internet surfing.

DVD Players and Blu-Ray Players

Recorded movie-player technology is in the process of taking a leap forward. Most homes have DVD players, and DVDs have become popular choices for buying movies and for recording home movies. You may have heard about the latest "format war" (remember VHS vs. Betamax?) involving HD DVD players and Blu-Ray players. The makers of HD DVD players have surrendered, and now **Blu-Ray** seems to be the format that will prevail in the next few years. **Blu-Ray** disks have more than five times the storage capability than traditional DVDs (25GB on a single-layer disk or 50GB on a dual-layer disk. This allows for a lot more information to be stored on a single disk, and some movie producers hope to use this to add interactive features.

With this new technology in the process of development, we would not suggest, at this point, investing in an expensive new DVD player.

Should you upgrade now? Here are some things to consider:

- According to Blu-Ray.com, most, if not all, Blu-Ray players will be backwards compatible, which means they will play DVDs. So you will not have to replace your DVD collection.
- As this is a new format, prices are still on the high side. But now that the standard has been accepted as the

new step-up from DVD, there should be more competition and lower prices.

Checklists

Are Your Existing TV Sets Digital-TV Ready?

TV #1

Manufacturer and model number:____________________

____

Serial number and date of purchase:_________________

____

Warranty term:_________________________________

Does your television have an ATSC tuner? Y/N (See page 72 for more info and check manufacturer's website.)

How does your TV receive television signals? (roof antenna, rabbit ears, cable box, satellite box): _________

____

If the TV uses an antenna for signals and is not equipped with an ATSC tuner, do you have a digital converter box? Y/N

If using an antenna, do you need to upgrade your antenna for better reception? Y/N

TV #2

Manufacturer and model number: ____________________

____

Serial number and date of purchase: __________________

____

Warranty term: ______________________________

Does your television have an ATSC tuner? Y/N (See page 72 for more info and check manufacturer's website.)

How does your TV receive television signals? (roof antenna, rabbit ears, cable box, satellite box): _________

____

If the TV uses an antenna for signals and is not equipped with an ATSC tuner, do you have a digital converter box? Y/N

If using an antenna, do you need to upgrade your antenna for better reception? Y/N

TV #3

Manufacturer and model number: ____________________

____

Serial number and date of purchase: __________________

____

Warranty term:________________________________

Does your television have an ATSC tuner? Y/N (See page 72 for more info and check manufacturer's website.)

How does your TV receive television signals? (roof antenna, rabbit ears, cable box, satellite box): ________

____

If the TV uses an antenna for signals and is not equipped with an ATSC tuner, do you have a digital converter box? Y/N

If using an antenna, do you need to upgrade your antenna for better reception? Y/N

Choosing a New TV

Size of room you are shopping for: ________________

Average distance between seating area and TV placement: ________________________________

Lighting condition (bright, average, low-light):________

____

New TV Comparison Chart

Type of TV	*Plasma*	*CRT*	*Other*	*Other*
Price				
Requires professional installation?				
Requires home delivery?				
Requires new stand/ cabinet?				
New connectors needed?				

Your New TV Records

Antenna required? Y/N

Manufacturer/Model number: ______________________

Serial number and date of purchase: ________________

Warranty term: _________________________________

Programming Content Provider Records

Provider name: _________________________________

Account number: _______________________________

Customer service number: _______________________

Monthly service fee: ____________________________

Contract term: _________________________________

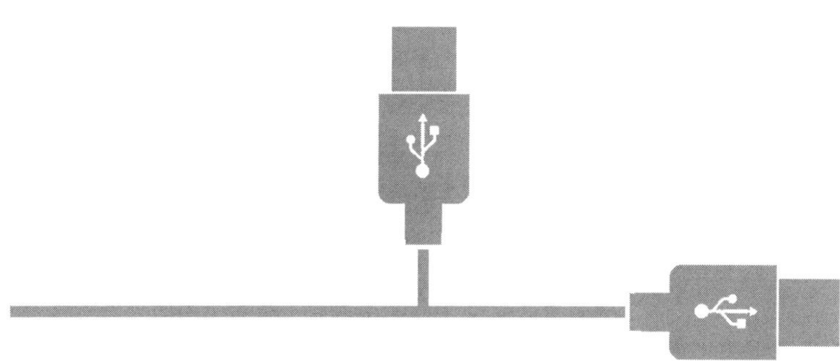

5 Cameras

In buying a camera, most users want an *affordable*, *easy-to-use* camera that takes pictures that will look as good as printed snapshots, on their computers, and in e-mails. A camera should be the right size so you feel comfortable bringing it with you on vacation and be able to work in a variety of lighting conditions. Most of these cameras are known as **compact digital cameras** or **point-and-shoot cameras**.

Sometimes you may want something more from your camera such as the ability to accept other lenses so you can get exceptional close-up or action shots. You may want to be able to take photos that can be used in professional situations—for flyers, brochures, or other commercial uses. You may want to use manual adjustments (as opposed to automatic) so you have greater creative control of your images. These uses might require a step up from a good digital camera.

If you want to become a camera hobbyist or take professional quality pictures, you would probably want to consider a **digital SLR** (single lens reflex). This means that there is a mirror and prism system to capture your images on the sensor. This allows for varying the **focal length** of the image and altering the **depth of field**. Think of those images you have seen where something in the foreground of a picture is super sharp and the background is blurry or out of focus. Digital SLRs or DSLRs often resemble professional film cameras with the ability to switch lenses and accept professional flashes.

In either case, you want to know what all of those abbreviations mean so you can make an informed choice. We'll walk you through the features and specifications so that when you go to the camera counter, you can be confident that you've chosen the camera right for you. We will then turn to your options to share, print, and publish your pictures.

Choosing Your Camera

Size and Shape

We address this first because if you are looking for a compact camera that will fit easily in your pocket, you won't want a bigger, bulkier camera even if it has many more features. Think of how you will use it. If you are looking for a camera to always have on hand, a small compact camera will suit most of your needs. If you are comfortable with a slightly larger camera or even a camera on a neck strap, you may like the more solid feeling of a larger camera.

It is important that you like how the camera *feels* in your hand. You should be able to read the menus easily. Your right index finger (sorry left-handed shoppers—left-handed digital cameras aren't readily available) should naturally go to the **shutter release** (the photo-taking button), and not block the flash. Can you easily see the picture's composition in your viewfinder? Do you prefer to have a preview screen? Many cameras will only have a **screen** (and no view finder) that serves the role of a preview screen, playback screen (to show you the picture you just took), and menu screen for showing you a menu of options for controlling pictures.

When you read a lot of product reviews, often the emphasis is on the technical specifications and advanced capabilities of the camera. It is important to treat the very basic issues we've just discussed—will this camera work for me?—as a first step, so that that expensive camera that you just bought is actually one you will use.

Resolution

There are two main considerations in making crisp photos: the number of **megapixels** (one megapixel = 1 million pixels) and the **image-sensor** size. Most shoppers are familiar with seeing cameras ranked by megapixels, but they aren't as familiar with the image-sensor size.

Megapixels—If you think back to the idea of each dot in an image being a pixel, you can understand that if a camera's image has more pixels per image, it will be sharper and show more detail. The more detailed the image, the larger you can make the prints or the size of your image on a computer screen before it starts to look fuzzy or **pixelated** (where the pixels composing your picture become noticeable).

If you are primarily taking snapshots for simple picture

sharing in e-mails and on the web, a two megapixel camera will yield high-quality snapshots. But if you want to make larger image prints or use the images for flyers or full-screen web shots, you will want a higher resolution camera. The flipside to this is that your camera's memory card (we'll get to that in a minute) will hold fewer higher resolution images, so you may wish to shoot everyday pictures at a lower resolution and use the higher resolution settings for more important events such as weddings and graduations.

Image-sensor size—This is somewhat equivalent to the size of film you used in nondigital cameras. Remember the old 110 film used with the original automatic cameras? That film produced okay little snapshots, but you couldn't blow up the pictures to quality 8x10s. The type of film used in 35mm photography is larger than that of 110 film, so it produces a larger image that retains most of its crispness when enlarged.

The sensor size follows a fairly complicated naming system, but the simplest way to put it is that a larger sensor size is better. Most digital cameras range from 1/1.8 to 4/3 inches. Cameras in the $200 range should have a sensor size of at least 1/2.5 inches. This specification may be listed as simply "type," referring to "type of sensor."

There are two main types of sensors: **CCD** (short for charge-coupled device) and **CMOS** (short for complementary metal-oxide-semiconductor). The technology behind each type of sensor is different, but at this point, one cannot be said to be superior. The important difference is sensor size. Most consumer cameras are CCD, so you are likely to see the type listed as 1/2.5-inch CCD.

I know this all sounds very confusing and technical, but you should understand why some cameras cost more and know how to compare between two similar cameras. Also, when you get to crop your photos (using software tools like photo editors to select sections of the image to cut off), you want to make sure that the image that is left after the crop is of high enough resolution to look good occupying the whole print.

Lens Type and Zoom

Most compact cameras do not have interchangeable lenses, but do provide some **optical zoom** adjustments. These allow you to zoom in to capture a larger image of a smaller object than if you just used the standard setting. A 3x zoom is pretty standard for mid-range cameras. To return to the 35mm example, it is the equivalent of a 38-105mm lens attachment. This will work for most users. If you need a more powerful zoom you will need to look for one of the higher end digital cameras, which allow you to use add-on lenses.

Generally, digital zoom is irrelevant. If you rely only on digital zoom, it means that the same image will just be cropped to make the smaller objects appear larger. If the image is large enough, digital zoom may not make the picture unacceptable, but it is not equivalent to using a larger lens to make that adjustment.

Controls and Menus

This is another subjective but important way to judge a digital camera. The menu of options (automatic flash v. no flash, red-eye reduction or not, zoom, size of the pictures) needs to make sense to you after you review the basics in your manual. The best and only way to figure this out is to try a hands-on test run in your retail store or to explore a friend's camera. The menu should not just be composed of symbols that aren't obvious to you. The button you use to navigate the menus should be easy for you to control and not subject to changing unintentionally by a slip of the finger.

Pay attention to shooting speed. Try to see how long the lag seems between pictures. Is there a long shutter lag (delay before you can take another picture)? Cameras have been improving this shot-to-shot speed—look for continuous shooting or burst mode as features. Try it yourself, though, and see if the speed is acceptable.

Battery Types and Recharging

Digital cameras can go through a lot of battery power. Alkaline batteries are convenient because they are available almost anywhere. Alkaline batteries include AA or AAA batteries that also work in your remote, clock radio, and other small gadgets. Unfortunately, these batteries don't usually provide a long life. It is best to rely on rechargeable batteries such as rechargeable lithium-ion or nickel-metal-hydride batteries. Any camera that uses rechargeable batteries should include a recharging station as part of the package, or a recharger should be available as an upgrade. Make sure to include the cost of a recharger, if you need to purchase one separately, when you compare prices for cameras. Most users will want to consider buying a second battery so that they are prepared for uninterrupted photo taking.

Flash Controls and White Balance

Most cameras will include a basic flash mechanism that can auto-detect when a flash is needed. You might also want **red-eye reduction**. This works by flashing many times, causing the pupils of those being photographed to contract—which in turn reduces the chance that eyes will look red in the final picture.

Some more sophisticated features available include adjustable illumination, so that you can achieve the desired level of lighting. Larger cameras may also have the capability to attach external flashes to your camera. If you are moving beyond a small pocket camera, make sure this is an option. These flashes use a separate battery, will not drain the camera battery as quickly, and are ready to use in faster time than a built-in flash. They also can provide greater illumination.

You may encounter the term "**white balance**." All digital cameras have an automatic white balance feature or **AWB**. It detects the **color temperature** (the differences between fluorescent, candlelight, sunny daylight, and twilight) of a setting and adjusts so that the color white is reproduced as white in your image regardless of the light source. Unfortunately, while the camera usually does a pretty good job, you may be able to achieve better results. Some cameras have settings for cloudy days, sunny days, and indoor or snowy days. This will further refine your pictures. Some cameras will allow for manual adjustment. You simply would hold a white card, or piece of paper, and adjust the camera's settings until it is showing the card or paper as being true white.

Storage Media and Connector Types

Most digital cameras have a small amount of internal memory but rely on some form of external memory. There are **CompactFlash**, **SD** (Secure Digital) and **SDHC** (Secure Digital High Capacity), **Sony Memory Sticks** (only required and useable by Sony Products), and **xD-Picture** memory cards. Less common media storage types include mini-CDs or other hard-disk storage. These all vary in price, size, and flexibility.

In general, the type of storage media shouldn't affect your decision, but memory cards that are only useable by your particular brand of camera are going to be more expensive and perhaps harder to find. It also is useful to have multiple devices that use the same card, so that you can share images easily.

As for connectors, some brands use special **docking stations** (which also usually serve as a battery charger) that connect to your computer via its USB connection. Others use a FireWire connector (somewhat rare). But most cameras connect via USB connectors. A more efficient way to connect is to use a **memory card reader** (simply a slot on some PCs or printers). Many printers can accept different types of memory cards, and dedicated photo printers will accept multiple card types.

One thing to be aware of is that some printer memory card readers are **read-only**. This means that you can upload your pictures from your camera's card to your printer or PC, but if you want to move pictures from one card to another or from your PC to a memory card (to take to a shop for printing, for example), a memory card reader that is read-only won't be able to **write** (in other words, to save) to your card.

We recommend considering a **memory card reader/ writer** that attaches to your computer through a USB connector. This affordable gadget allows for the most flexibility in using these handy cards. Often this is a much quicker and less battery-draining alternative than using your camera's upload feature.

Digital Compression

Your digital camera will have a setting for the level of **digital compression** you want when saving your photos. Your fancy new camera will have more digital information available to save then you could see when printed in a smaller format such as a 4x6 print. Your camera can compress the image it takes by scanning for duplicated information such as colors in a background.

Most digital camera users want to be able to take as many pictures as possible with their camera and memory card and be able to send these images to friends, share on the web, and print from time to time. To save storage space, most users will set their compression level fairly high (meaning more compressed), but in doing that you are losing the very detail that you paid extra for when it came to buying more megapixels and larger sensor size.

Keep in mind that *you can never uncompress an image once it is saved.* There are many reasons to compress the final images (so you can share via e-mail, take up less room on your computer), but perhaps consider saving your images as less compressed when you anticipate that you would want a larger print. You can always make your images smaller using various software tools, so we suggest using a higher-quality setting when you use your camera if you are unhappy with the quality of your prints.

Most cameras will have a setting for photo quality; you will get the best results with the higher or best setting but there will be tradeoffs. The photo file size will increase, it may even double from the standard setting, so you will fill up your memory card quicker. But as memory cards are getting bigger and cheaper all the time this should soon not be a big problem. Larger file sizes will also increase the time between taking photos on some cameras. The chart below is a rough estimate of file sizes. Depending on the camera and the file formats, these can vary greatly, but it is meant to provide a rough idea.

	Good	*Better*	*Best*
Maximum Print Size (in inches)	5x7	8x10	20x30
Size of File (approx. in megapixels)	1 MP	2 MP	4 MP
Storage Space (approx. in MBs)	3 MB	6 MB	12 MB

Other Features to Consider

Image Stabilization

For someone who cannot hold a camera steady, especially in low-light situations, this is a must-have feature. Shaky hands have been ruining pictures long before digital cameras, and if you do not want to carry a tripod everywhere, this feature can compensate for a little too much caffeine! Of course, the investment in a tripod will go a long way towards much improved photographs.

Face Detection

This feature looks for a face or faces in a potential shot and adjusts to focus on a face during the camera's auto-focus process. Your eye will be a better judge, but it can be a neat feature for quick and effortless photo taking. There are limitations—such as the face must be looking directly at the camera. Or it may choose a face in the image instead of the object you were intending to focus on. So be sure to know how to turn this feature on and off depending on the situation.

In-Camera Editing

Some basic picture editing can be done on cameras equipped with this feature. The advantage here is that you can correct the image right when you take it instead of having to remember to edit at a future time.

- **Remote control.** A nice feature for portrait taking. Most remotes for digital cameras are small and wireless, so they might be impractical for users who don't want to carry camera accessories.
- **Video or other multimedia recording.** Some cameras will have the gee-whiz feature of video- and sound-recording capabilities. This can be useful for short little videos, but will be less than perfect when compared to a dedicated video camera. You can compare it to early cameras on cell phones. In a few years, you may be able to combine these two devices, but not quite yet. Still—this

neat trick can be a deciding factor between two similar cameras, and a short video is often the perfect size for sharing with friends. Not all can record sound, and some limit your film time to thirty seconds.

- **Web cam.** One way to expand your mobile office is to use your digital camera as a web cam for videoconferencing, when used in conjunction with your computer. If you want to use this feature, be sure to check to see if this also supports sound.
- **Wireless image transfer.** This feature allows you to send your pictures back to your computer without a physical connection.

Photo Editing, Storage, and Printing

Photo Editing

Once you have uploaded your pictures to your computer, you will find that this can be the beginning of truly enjoying digital photography. Photo editing software means that you can touch up, crop, and add effects to your photos in ways that allow you to be a creative genius. Your camera will have basic software, with some offering a wide-range of features and a proprietary photo-printing store. These may be convenient and easy for you to use, but don't think that these are your only options for photo printing, editing, and sharing.

Professional suites such as **Photoshop** from Adobe are powerful and expensive. They are the industry standard that has familiarized us with the trademarked verb "to photoshop," undoubtedly the true "diet and beauty secrets" of the rich and famous. The good news for us nonprofessionals is that these pro-level tools are now rivaled by free software such as **Photobucket** (from Google) and **GIMP** (an open-source program).

You should explore the software that may come packaged with your camera or other free editing suites and see which is easiest for YOU to use. Don't rely on someone saying that this one or the other is the best—these are all powerful tools, and you should choose the one that works the best for you. You can read reviews of popular software downloads at CNET's Download.com.

Photo Storage

One of the main advantages of digital photography inevitably leads to one of the main disadvantages of digital photography. We can take hundreds of pictures, since once we buy the hardware and memory card the pictures are "free." This allows you to take lots of photos, so that the one-in-a-million shot becomes much more achievable if the less-wonderful pictures can be cleared with a click.

But what happens if your computer fails? Or you accidentally erase them from your computer's hard drive? Or the pictures reside on your memory card in your camera—never uploaded to your computer—and then the camera is lost, stolen, or damaged? After every natural disaster you see family members mourning the loss of their precious family photos, which goes to show you how important these snapshots are to us. Digital photos should be treated as precious objects as well.

There are two ways of preserving your photos. One is to backup your computer onto an external hard drive or on one of the online web storage sites. We highly recommend a good external hard drive. An affordable external hard drive can store thousands of photos, your video files, music, and data files.

Photo-Sharing Sites

These sites allow you to post pictures on the web to share either with the entire Internet or with selected friends and family. Picasa and Flickr are the most popular Internet-based tools. Some photo printing sites such as Kodak.com and Shutterfly.com allow you to create galleries for sharing and printing.

Photo Printing

You have many options for printing. You can still go to your local photo printer wherever they may be—in a camera shop, drug store, department store, etc. Some will allow you to upload your pictures through their website and then pick up your photos in the store. You can also bring in your memory card, or a CD or DVD, with your images on it. Self-help kiosks are also available in many stores. Each of these options allows you to print high-quality prints without your own printer, ink, or photo paper.

Home or Portable Photo Printers

These are popular and have become quite affordable. Some portable printers can work without your computer for on-the-go printing for use at parties or meetings. Unfortunately with personal printers, the price per print can be quite high. Be sure that you factor that cost (which can run about 25 cents a print and up for the ink and paper alone) when deciding whether to print from a dedicated photo printer.

To balance the convenience of printing photos at home with the high cost, you might use your desktop printer for the occasional print or two and use professional printing for bulk copies. If you do purchase a dedicated photo printer, one feature to look for is the ability to accept direct printing from your memory card. Also, you may have to use the dedicated software for your printer—read the instructions.

Digital Frames

One accessory that is becoming very popular are digital frames. You can display a few or a whole library of photos in one single frame. Some digital frames can receive images wirelessly on your home network; others rely on a memory card. Some frames will also play music, but this feature is probably one you can do without.

Camcorders

Many of the specs and terms for video recording are the same for still photography. There are a few others you need to know, but the basics remain the same.

Design and Weight

Make sure that you test-drive a camcorder before committing. While you may not find one that fits in your pocket, with the move away from VHS cassettes, cameras have gotten quite small and lightweight. Be sure that you can operate the controls, that it feels good in your hands, and that the use of the camera seems natural. If you invest in a camera, you want to be sure that it is comfortable enough to use.

Type of Recording Media

For serious movie making, or to allow for computer editing of your video, you probably want to record to 8mm or Hi-8 tapes in the **DV format**. For playback, you will have to have either the computer tools necessary to export onto a DVD or use your camcorder as a way to play the video through your TV.

Many home movie makers prefer to record directly to a DVD-R or DVD-RAM disk in the **MPEG-2 format**, which can be played back on most DVD players or computers. With the most compact camcorders, many record on internal flash memory or memory cards and store videos in the **MPEG-4** or **MPEG-1 format**. These formats aren't as high quality as the DVD or MPEG-2 formats, but they are used for making the smaller videos popular on Internet video sites such as YouTube.

Lens

You should look for at least 10x optical zoom. Again, digital zoom does not increase performance. Higher-end models will allow you to add lenses for distance or special effects.

Image Stabilization

With video recording, image stabilization is more critical than with photo taking. A shaky video can be unwatchable. With a larger camera at formal events, a tripod is still probably required. However, technical advances in both **optical**

image stabilization and **digital image stabilization** have greatly improved the smaller, hand-held models. Now even movies shot-on-the-go are leaps better than the horror-movie shaky-screen effect that was previously the norm. Optical stabilization is the preferred technology, but it can be expensive.

Resolution and Sensor Size

More pixels and larger sensors yield crisper pictures. Higher-end cameras will have more than one—usually three sensors. The pixel rating will be per sensor. Expect to see ratings of 680,000 pixels and up per sensor and sensor sizes of $1/6^{th}$ inch and larger.

Illumination Rating

This measures the ability to film at lighting levels lower than full daylight. This is usually measured in terms of **lux**, which measures the level of light. The lower the lux rating, the better. Expect to see ratings of 2 to 7 lux. For perspective, moonlight is generally considered to be 1 lux, family living room is 50 lux, and typical outside daylight is 32,000 lux.

Progressive Scan or HD Capability

Higher-end cameras (about $800 and up) should be able to film in HD quality or Progressive Scan.

Other Features

Other features and specs to pay attention to include microphone jacks (plug), a headphone jack, flash attachments, type of connectors, basic software for video editing, and still photo settings.

Checklists

Camera Comparison Chart

Camera Name			
Type (point-and-shoot or compact)			
Size			
Megapixels			
Image-sensor size			
Optical zoom			
Other features:			

Your Digital Camera Records

Manufacturer model and number: ____________________

Serial number and date of purchase: ________________

Warranty term: _________________________________

What media-storage types can it use? _______________

Media type purchased: ___________________________

Photo-sharing site's URL for your pictures: ___________

Camcorder Comparison Chart

Name			
Weight			
Media used			
Optical zoom			
Image stabilization			
Resolution and sensor size			
Illumination rating			
Progressive or HD			
Other features:			

Your Camcorder Records

Manufacturer model and number: ____________________

Serial number and date of purchase: __________________

Warranty term: ______________________________________

What media-storage types can it use? _________________

Media type purchased: ______________________________

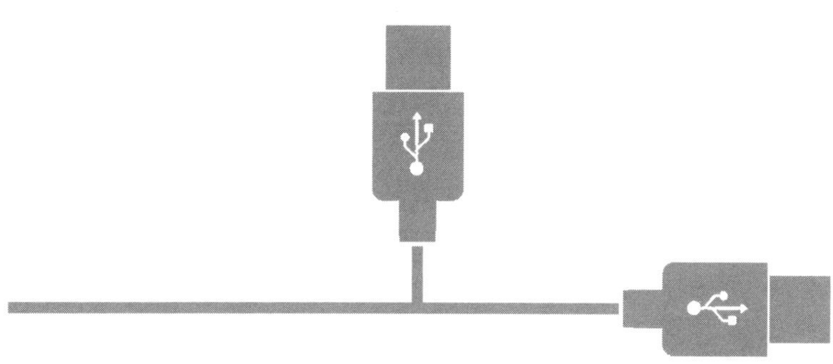

6 Media

The digital revolution has brought great changes to how we acquire and use different media—music, video, photos, and even the written word.

One of the greatest changes brought about by the Internet is how we "consume" the written word. Anyone can put his or her thoughts and opinions online for the world to see by starting a **blog** (web log). Another significant change is how we read the news and even books. Every major paper has a website that updates the news as it happens and makes it accessible to readers around the world.

Millions of **electronic books** can be downloaded, along with the daily paper, to small handheld devices such as the **Sony Reader** or the **Amazon Kindle**, or to your desktop or laptop computer. While this is all still fairly new, the possibilities are extremely exciting and the advances are changing the way we get our information and entertainment.

Another major change involves how we access music and video. Small **portable media players (PMPs)** allow you to carry in the palm of your hand an entire music collection that would have filled rooms with CDs or LPs. Imagine hundreds or thousands of songs filed in an easy-to-sort, organized manner. Most of these devices can also play videos, movies, and TV shows available for download from online stores.

Where to Begin

Learning about both the basics of any device and your options is important before you commit yourself to a purchase. This process will help you consider how you would like to use your media collection. It is important, as well, to be aware of compatibility issues regarding the various devices and files. Some file types, for instance, won't play on certain manufacturer's products. Paying attention at the pre-purchase stage will save you set-up time down the road and will ensure that the media you buy will work on the "players" you've purchased.

Everything Working Together

In an ideal world, you would have a PC that has all your music and video files in one place, and you could listen to them in any room in the house. You could play your videos on any PC or TV in the house, as well as put them on a portable device and take them with you to have in the car or on a flight. As this is the ideal situation, we will work through the step-by-step process of how you can achieve this level of flexibility and portability. You don't have to do this all at once. Go piece by piece and choose what options work for you.

Music

Do you have a collection of CDs? You can start your digital music collection here by converting them to files that will work with your new media world. To do this you will use a program to convert the music on the CD to a file that will work on your home media system. In effect you are **uploading** your CDs onto your computer or **ripping**—reading and storing the information on your CD to your computer. This will allow you to export or **download** this information to another device or computer or **burn** (copy) files from your computer to something else like a CD.

If you have a Windows PC, a good program to use is **Windows Media Player**, and it comes free with your Windows Operating System or via a free download from Microsoft. If you have an Apple computer, the tool you will most likely use is **iTunes**. This is also free when you buy your computer or free via a download from Apple. Both programs will work on the other type of operating system, but the features for each are optimized for their "native" system.

Media Types

All of the file types have advantage and disadvantages, but the most important thing to keep in mind is sound quality. Several things affect this. First is how the music is **encoded** (this is the way the file is converted from a large file on a CD to a smaller file that is more portable or PC friendly). It is sort of like recording an LP (remember those?) to a cassette tape. The cassette tape is smaller and easy to carry around but never sounded as good as the LP.

In modern conversions, the drop in sound quality—while it does exist—is not very noticeable, and most people won't hear the difference at all if you do the converting correctly. The second thing that affects the sound quality is what you play it back on—the device itself and the speakers. A big, high-end stereo with big, expensive speakers will usually sound much better than a small pair of inexpensive headphones. The trade off is that the loss in sound quality is balanced by the portability of small devices.

The last item is something called **bit rate** or **sample rate** (the amount of information copied from the larger source—your CD—to the new file). The rule of thumb here is a higher bit or sample rate equals better sound. Higher bit rates also require more storage space, but as storage space is getting cheaper all the time, it's not as big a factor as it used to be. Let's now address the options for each of the file types, the pros and cons of each, and how to change the default settings for the best quality.

Throughout this book, we have avoided the exact "how-to's" on the particular software or hardware discussed, as there are so many variations and changes happening all the time. Nonetheless, as there are two main players in the media market, and it is so essential to the discussion, we want to include a brief discussion on choosing media types in Windows Media Player and iTunes.

Settings in Windows Media Player

With your mouse, right click on the bar at the top of the application. You will get a menu from which you can select **Tools**, then **Options**. A window will open with several tabs on the top—choose the **Rip Music** tab. There are two sections on this page—one section is where the files will be stored; the other section in which we will be working is called **Rip Settings**. Under the word **Format**, you will see a box where you can choose the format you want.

The default setting is **Windows Media Audio** (WMA—discussed below), and there are several other choices. The two to choose between are the default—WMA—and MP3. We don't recommend choosing any of the others as they don't work as well with other devices. The other option to look at is **Audio Quality** (which is the bit rate or sample rate mentioned above). There is a slider bar where you can move the arrow from **Smallest Size** to **Best Quality**, and when you move the bar, it will show you how much space a CD will take up on your PC. We recommend moving this slider all the way to **Best Quality**, then click OK when you have finished making these changes.

Settings in iTunes

From the **Edit** menu choose **Preferences**. A window will open with several tabs along the top—choose the **Advanced** tab. You will have three sub-tabs to choose from—select the **Importing** tab. Next you will see the options for importing CDs in the **Import Using** box. The default is **AAC Encoder**, and the recording quality setting is defaulted to **High Quality** (128kbps-128 kilobytes per second). For greatest flexibility, we recommend you consider changing this to MP3 Encoder. In the recording quality settings, we recommend choosing **Custom**. A new window will open in the **Stereo Bit Rate**—choose **320kbps**.

Both of these programs are great for managing and importing your music collection, but there are some things to keep in mind. iTunes is designed to work with iPod music and video players ONLY, and many features of this program require you to have an iTunes account, which requires a credit card to set up. One convenient feature is that the online store is built into the program, so buying new music is a breeze—but only for iPods. Windows Media Player's main disadvantage is it will not work with iPods, but it does work with all other players.

Why Convert Them at All?

Files on a CD are in a format called Wave or WAV files. These are quite large and would take up a lot of space on your PC—an average about 600 to 700MB. By converting your CDs when you rip them, you will save a lot of space.

Windows Media Audio (WMA)

This is the file type most associated with Windows-based PCs and will work on all media players except iPod. A converted CD will only take up about 50 to 70MB with the default quality settings, and 80 to 100MB on the best quality setting.

Pros:

- Very good sound quality
- Small size

Cons:

- Will not work in an Apple iPod

Advanced Audio Coding (AAC)

This is the file type most associated with Apple products such as the iPod. Older versions of iTunes converted a CD to only this file type. This is also the type of file you will get when you buy **copy-protected** music (files that have software included to prevent unauthorized copying of copyrighted material) from the iTunes online store.

M4A

This is the file type iTunes now creates when **ripping** CDs. This file type is supported mainly by the iPod, but some game consoles will now play these files. This is also the type of file you will get from the iTunes online store for files that have no copy-protection.

With both of these, a converted CD will take up about 40 to 60MB with the default quality settings, and 70 to 90MB on the best quality setting.

Pros:

- Very good sound quality
- Small size

Cons:

- Will not work on most portable media players

MP3

This is the most widely supported file type. All portables will play this file type, so it is a good choice if you are going to mix Apple and Microsoft systems together. But if you are going to be using one system only, we would recommend the WMA for Windows systems, and AAC or M4A for Apple systems. The main reason for this is there is a major limitation in the MP3 format: it has a hard time converting music in high frequency (higher notes) range. A converted CD using MP3 will take up 50 to 70MB on the lowest setting and 140 to 180MB on the best setting.

Most audio files without copy-protection can be saved as MP3s as well, which is a great option for burning CDs to be played on regular CD players. Most newer CD players will play these CDs.

Pros:

- Widely supported

Cons:

- Not the best sound quality

Video Files

Video files should be just as easy to use but worries about illegal file copying have made it extremely difficult to just pop a DVD into your PC and convert them to a format that can be played on other devices. Most commercial videos (as in the movies you purchase) have a layer of copy-protection called **CSS** (Content Scramble System) that prevents copying. There are products available that can break this copy-protection, but it has been found to be illegal to circumvent these protections, even if most copyright experts believe that a copy for personal use is acceptable under fair-use doctrines.

There are legal ways to get video content, however. Apple sells video content through its online store—both TV shows and movies. Amazon.com sells and rents TV shows and movies as well, and more sites are opening all the time where you can buy legal copies of video content.

Taking It with You: Portable Media Players (PMPs)

MP3 players, Zunes, iPods, PSPs—whatever you call them—have become a great way to keep your music, photos, and video files all in one place and to share them with friends and family. The biggest considerations are storage space and design. Are you going to just have a few music files to listen to while you go to the gym, or do you want to larger collection of music and video for long flights? And just as importantly, do you like the way it looks and how the controls and menus work?

Storage Size and Type

There are two main types of storage in PMPs—**flash memory** and **hard drive**. Flash memory refers to small memory chips that can hold your files even if the power is off or the battery is dead. There are no moving parts, so they should last a very long time and are extremely durable. The main drawback is the cost—it's expensive to get enough space on them—but this is changing quickly. For now, hard-drive PMPs have way more storage space for the money. The main advantages of hard-drive storage are price and size—they are bigger and you can store more. When you want to use video files, which are larger than music files (500 to 600MB for a typical thirty-minute TV episode and around 2GB for a full-length movie), this is important.

Amount of Storage

This is indicated in **GB** (or gigabytes, 1 GB = 1 billion bytes). Keep in mind the average feature-length movie download is about 500MB (megabytes, 1 MB = 1 million bytes) or half a GB up to 2GB for DVD-quality. A TV show is about 200 to 600MB. An average song is about 2MB. So if you plan on storing a lot of stuff on your portable device, get one with a lot of storage. It might be more practical to store media on your PC and put them onto your PMP as needed.

Buying Music for Your PMP Online

Many online stores sell music for portable players. The iTunes music store is great for iPod users, but not good at all if you have another type of player, as all of their stuff will only work on an iPod. Amazon has a good selection of music that will work on all players if you buy the files in MP3 format. The online Wal-Mart has a great selection of MP3 files as well. At both of these online stores, you will also find a large selection of WMA files that will not work on the iPod.

Buying Video for Your PMP

Apples' iTunes online store has a great selection of TV shows and movies that you can rent or buy, but they will only work on an iPod. Amazon's Unbox store also has a great selection of movies and TV shows that you can rent or buy for your PMP as long as it supports the Play for Sure standard. Almost all new non-iPod PMPs are on this list. You can look for the Play for Sure logo or Certified for Windows Vista logo when purchasing or go to www.playforsure.com to see if your device is on the supported list.

Using Your PMP in the Car Stereo

Many factory-installed car CD players will play MP3s burned onto a blank CD. Another feature that is becoming standard in cars is the ability to plug your iPod or other PMP into the stereo. This is a big step up from the previous option of using car-conversion kits that relied on your cassette deck (not even available in most newer cars) or a radio transmitter (which was subject to interference from radio broadcasts or other car stereos).

Using Your Media Files in New Ways

Imagine having a slide show of family photos on your TV, maybe with a custom sound track of your favorite music playing in the background. Or having the video of the family vacation play on the TV through the home network, or having the music stored on your PC play and be controlled in another room.

Using your PC as a central storage place for all your media files and being able to share them across your home network is becoming easier to do these days. Photos, music, and video and can be played back on different TVs or on your stereo or even through your gaming system, allowing for a truly networked home.

This is all possible by using a **streaming media device**. Streaming media refers to the ability to send media files—music, video, or photos—stored on one networked device such as your PC to another networked device. The device doing the playback usually does not require a way to store the media file on its own; it takes the media from other devices and streams it across the network as needed.

Streaming Media Devices

There are several ways to do this. You could get a device dedicated to performing this task. Sometimes these are referred to as a **streaming media server**. These hook into your TV and stereo (either wirelessly or through cables) and use your home network to access the files on your PC. They have a remote control and are extremely simple to set up. You can then access your files stored on your PC through the TV and play them back.

What to Keep in Mind

Streaming media devices will have a list of media file types they support. For example, Apple TV aims to be a full-fledged media server, but it only supports Apple media. A lot of research is required here if you already have a large selection of media files on your home network. If you are just starting out, however, many of them come with easy-to-use tools to convert and set up your media files to work seamlessly with the device.

Drawbacks

In order to use streaming media devices, you will need a fast network, and sometimes this cannot be achieved using wireless home networks alone, especially if you are streaming High-Def video files. Your best bet is to use a wired network connection for video files, but for music and photos, a properly configured wireless network will work just fine.

Video Game Consoles

Some home gaming systems can also be used to stream

media across your home network. Microsoft's **Xbox 360** and Sony's **PS3** will do this and will also play games and DVDs. The Sony PS3 also includes the new **Blu-Ray DVD** on it as well. These devices are designed to be extremely easy to set up and use on your home network. They also have shopping services where you can download games, TV shows, and movies in High Definition, and soon they will have music stores and the ability to transfer these files to a portable device, hopefully eliminating the need for a PC. The main drawback is they have limited storage capacity, and the content downloaded from the online stores is not playable on other devices.

Video Game Consoles for Gaming

The makers of video game consoles are working hard to attract people that normally would never use or play a video game. The normal shoot-and-blow-things-up type of game, while still extremely popular, is not the only thing out there. Many new games are being made for the whole family, and new game controllers make learning these games extremely easy. With Wii Sports, if you know how to swing a tennis racket, then you can play a game of tennis by swinging the game controller the way you would swing a racket in real life, the same with golf or bowling and many other games. You can re-form your old rock band with the game Rock Band by using guitar-shaped controllers. And unlike games in the "real" world, you can play online with someone half way around the world. There are Jeopardy-type quiz games, card games, and many more. Quite a few are designed with the whole family in mind. If you think of them as you would board games, the concept of "family gaming night" takes on a whole new meaning.

Not only can you play video games on them, you can also download music, TV shows, and the newest movies in High Definition. They can also connect to your home PC and play your video music and photos as well.

Here's what to consider when purchasing devices and games.

Console Type

There are three major video game consoles on the market right now: Microsoft's **Xbox360**, Sony's **PS3** (for PlayStation 3), and Nintendo's **Wii**. We will go over the pluses and minuses of each, with a major eye toward the non-gamer, an emphasis on use of video game consoles as a central media device, and with the idea that these are fun devices to have and enjoy.

Game Selection

For the traditional gamer, this is the single most important decision factor. Each maker's stable of game franchises—Xbox's Halo series, Sony's Final Fantasy, and Nintendo's Mario universe—appeals to a very loyal fan-base. You will simply never see one of these brand-identified games developed for another game system. Beyond these games, most game developers will develop a version of the three major gaming systems.

Online Stores

All three of the game systems we will discuss have an online store where you can download content. They are not all the same, and the content will not work on any device except for the game console. They all have limited storage, but all systems also have removable storage (hard drives for Xbox360 and PS3, SD memory cards for Wii) that can be purchased as an accessory.

Motion Control

Nintendo introduced something called "**motion control**" to the video game world with its **Wii** game system. It uses special wireless controllers that have a motion detector in them that sends a signal to an infrared sensor bar. Therefore, it doesn't rely just on pushing buttons, but the very way it is held is registered by the game. This has been part of the appeal of the Wii—that feature makes it more intuitive to control what happens in the game through familiar motions such as swinging a baseball bat or a golf club. Sony and Microsoft have seen how this improves the gaming experience and are developing games and controllers that can work in this fashion.

Online Play

All three systems have online play, but they are all very different from each other. The main idea is to allow you to play online with people from all over the world in real time. You could join a race online with sixteen other people or work together to accomplish goals. This makes the game much more exciting because you are playing against other people instead of a computer.

Microsoft Xbox 360

Within the newest generation of game consoles, the

Xbox 360 has been on the market the longest. It has a wide selection of games that the whole family can enjoy, and a great online store for accessing the game system and for purchasing games and High Definition movies and TV shows. It also acts as a DVD player. As it comes from Microsoft, it works well with other PCs and will connect to your home network and play your media files with very little configuration.

Game Selection

Xbox has made its name as the home of the online experience of Halo.

Online Store

Of the online gaming stores, Xbox Live Marketplace has the most content (games, game demos, movie trailers, music videos, TV shows, and movies). Purchases are made via a system of "Microsoft Points" that you can purchase online or through refill cards sold by major retailers.

Motion Control

As of this writing, Microsoft is about to add a new controller and some games with this ability.

Online Play

Microsoft has the best online play there is right now, but it is not free. You have to purchase a membership for about $50 a year. Almost all games sold for the Xbox 360 have a significant online component with a very large community already on board.

Sony PS3

While a great system, this is geared more toward the hardcore gamer as it does not have as many family-friendly games. A big advantage is Sony's game console will play not just DVDs, but Blu-Ray DVDs as well. It's probably the most affordable way to buy a Blu-Ray player when you consider you get a high-powered gaming system as well. Sony is working toward this being a good home media system, which seems to be logical as they are a successful audio-and-video electronics developer, but as of this writing, it is not yet available in the U.S.

Game Selection

Sony's PS3 is home to the Grand Theft Auto franchise and Metal Gear Solid.

Online Store

The PS3 store is not as advanced as the Xbox Live Marketplace. It has game demos, games, and movie-and-game trailers in HD. Again, expect this to develop into a much more diverse online shop in the near future.

Motion Control

This was added late in the development of the system, not many games use it yet.

Online Play

Many games for this system come with online play, and one advantage over the Xbox is that online play does not require an annual subscription.

Nintendo's Wii

This is a great gaming system for families and has the most games geared for family and kids. It has none of the media-center options of the other two systems. It does have an Internet channel, a global weather channel, and a news channel. One sign of future capabilities is that in the United Kingdom, there is also a channel for the BBC's iPlayer channel—a download service for music and video content.

Game Selection

Nintendo is known across its gaming systems for the Mario universe, the Zelda adventures, and Metroid titles. Nintendo is known for its stable of family friendly games.

Online Store

The Wii has an easy-to-use online store that currently only sells games.

Motion Control

This has been one of the great achievements of the Wii system. Games such as Wii Sports and Wii Fit have started a revolution in how games are perceived and played by a public beyond the typical video gamer.

Online Play

Few Wii games have this option but more are coming. It is free to use but you have to know someone's "Friend Code" in order to play online with them. While this may protect kids from strangers online, it makes it hard to use unless you know other people who have the same games and system as you.

Other Ways to Access Media

Whether in your car, on the go, or at home, there are several new ways to access over-the-air media, providing a wide selection of music, news, and other information.

HD Radio

HD radio is a new and exciting way for radio broadcasts to offer higher-quality sound and more information. This is a new standard allowed by the FCC in the U.S. for broadcasting digital radio alongside traditional radio broadcasts. Special HD radio receivers are required, but these are becoming widely available. The stations operate as standard radio broadcasts; they are free, but usually supported by advertisements and donations. The HD content is found on channels associated with the frequency of the standard station. For example, 99 FM might have two or three channels—99 FM Channel One, 99 Channel Two, etc.—all in CD quality sound.

Satellite Radio

As of this writing, there are two major satellite radio providers, Sirius and XM, but they are merging into one company. This is a pay service where you pay around $12 a month for 150 music, news, traffic, weather, talk, entertainment, and sports channels. Many music channels play a specific type of music, such as music from the '50s or modern jazz, and most are commercial free.

The great thing about this type of radio is you can listen to the same station all across the U.S. and Canada without having to search the band for a new station to listen to—this makes it great to have on long drives. You can lose the signal for short periods of time when you drive through a tunnel or under a thick overgrowth of trees. A wide variety of receivers are available—for your car and home. They even have portable units that can also hold MP3s.

Checklists

Choosing a Personal Media Player

Do you have a large collection of digital music? Y/N

If yes, what format is it stored in (WMA, AAC, MP3s):

____

If mostly in AACs—then you probably want a player made by Apple Corporation. If MP3s or WMAs, you can choose any media player.

Media Player Comparison Chart

Name			
Storage size			
Type of storage (flash or hard drive?)			
Removable storage Y/N If removable, what types?			
Video screen Y/N			
If video screen, what size?			
Other features:			

Your Personal Media Player Records

Manufacturer and model number: ____________________

Serial number and date of purchase: __________________

Warranty term: _____________________________________

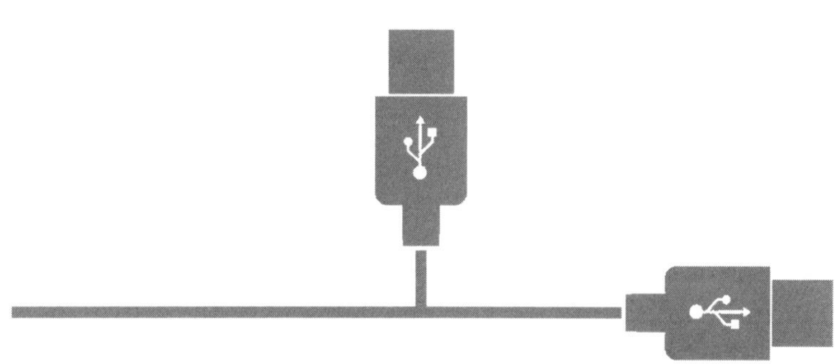

7 When Things Go Wrong

Things stop working. This is just part of life, but being prepared to deal with problems or having a basic understanding of how the "broken thing" works can go a long way to protecting yourself from an unskilled or unscrupulous repair person. In the previous chapters, we have tried to teach some of the basics of how your digital equipment works. In this chapter, we are going to shift gears. This chapter is a stand-alone tool on how to start troubleshooting the problem yourself. If that fails, we provide tips for hiring a good repair technician and for determining if the diagnosis sounds reasonable or if you are about to be ripped off.

Remember—organize before "disaster" strikes.

Organize Your Digital Life

A little organization can go a long way to keep problems at bay and to help you find what you need when problems strike. We recommend:

- Keeping one drawer or file cabinet just for warranties, manuals, and installation disks.

- Stapling your receipt to the manuals—this will be useful to prove your date of purchase for warranty purposes.
- Registering your products when you set them up. This is especially true with your operating system's installation disks. This, and other software you purchase, will have a "**key**"—usually a series of number and letters that indicate that you have a legal copy of the software.

It's also good to know what a device looks and sounds like when it is working properly. Notice the start-up sounds. Look at the lights that turn on and off as it is starting up. Does it have a regular hum? Does it emit a regular series of sounds? What lights are on when it's working properly? This will give you an idea when something is about to go wrong—for example, a sudden horrible mechanical "squeal," grinding noises, or blinking lights. If you start to see or hear different sounds—stop and backup your files.

Armed with a basic understanding of your digital devices, you will be better able to diagnose a problem or provide good information to a repair technician when something goes wrong.

Two Quick "Fixes" for All of Your Electronics

Tip 1. Shut it off, and wait for two minutes, then start it back up.

Many modern devices will freeze up and just stop working, often for no apparent reason. The general rule of thumb is to turn the power off for about two minutes, and, if possible, remove the power source—either disconnect the plug or remove the battery.

After shutting it down and waiting for a few minutes, power it back on and watch it do its normal start-up routines. Do all of the start-up screens look okay? Are there any warning messages? If there are warning messages, write the messages down. Some devices will not have any start-up screens, if that is the case, listen to it. Does it sound normal or are there odd buzzing or clicking sounds? Is there any abnormal beeping? If so, write it down. This will be of great help in finding and fixing the problem or problems.

If it starts up fine, that's great. These things happen once in a while. But if it starts to happen regularly, try to look for patterns. Were you doing the same thing that caused it to freeze up before? This is a good time to look through the device's manual and read the troubleshooting tips. This also may be a reminder that your device may be overworked—consider saving energy and the lifespan of your technology by turning it off from time to time.

Tip 2. Check the Connections.

This is a tip even the most seasoned pros will sometimes overlook. You may think that you have not touched any of the cable connections or the power cord, or that you never touch the battery, so how could that be causing the problem? Trust us, it happens. Cables come out, power cords come out, batteries don't sit in the holder quite right; it serves you well to just check them. We even go as far as unplugging the cable and plugging it back in. It won't hurt, and if it fixes the problem, then great. You would be surprised how often this works, and you can save yourself the expense and effort of getting it repaired.

Computer-Specific Tips

When things go wrong with your PC, not only is it extremely frustrating, it can be frightening as you may have all of your work and personal files on it. This is a good time to review some simple things you can do so when there is a problem or when something does go wrong, you will be prepared.

Disaster Prep

Computer hard drives fail. It has moving parts (the disk spins to read the data stored there), and things with moving parts fail. This is a fact of life, and it is the most common failure most PC owners will have.You need to be prepared. **Backup** your files on a regular schedule. Keep all the disks that come with your PC and any programs or game disks you may have purchased in one place. If there are any installation keys, don't lose them.

What should you backup? You don't need to backup everything on your computer; only backup the files you create—photos, documents, etc. All the other stuff such as the programs you run will need to be reinstalled, so you do not have to back them up.

Do you need to backup every day? That depends on what you do on your PC and how important what you worked on that day is to you. For most people once a week should be good but if you just finished a big project or just uploaded a bunch of photos and you don't want to lose all of your hard work, you can do a special backup. There are many automated backup programs that you can use.

Where do you put the backup? Some of the best backup devices are those small USB backup keys; or maybe an external USB hard drive—it all depends on the amount of data you need to back up. Remember, photo, music, and video files are large. You will need the appropriate storage space for those. Another approach is to copy your files onto a CD or DVD.

Keep It Cool

Computers generate a lot of heat, but they cannot perform well in overheated environments. Larger computer servers often have their own air-conditioning systems to make sure the machines are kept at an optimal temperature of 65-70 degrees Fahrenheit. Home users do not need to go to these extremes, but there are a few things you can do:

- Keep the vents of your computer unblocked and clean.
- Don't set up your computer in an area that gets direct sunlight.
- Don't set up your computer near heating vents.
- Keep several inches between your computer and any walls or other surfaces.

If a computer overheats, it will cause the computer's systems to malfunction and even corrupt your files. If you hear any squeaking, scratching, or other strange noises—it is often the fan malfunctioning. Pay attention to this and get it repaired—it can save your computer's life. Some "power users," like gamers, will even add another fan to their computer to make sure the machine operates at its best.

Keep It Clean—Literally and Digitally

Tidy up your workstation. Periodically clean your keyboard and mouse. Build-ups of dirt and grease can interfere with your system's ability to receive proper instructions from your peripherals. It also is a matter of sanitation. You may have seen articles comparing most computer keyboards' bacteria levels to those of toilet seats—take our word for it, have a regular system of cleaning with antibacterial wipes and compressed air. Never use any liquid cleaners, as you could damage the electronics.

Take out the trash. Scheduling maintenance functions on your computer will result in proper data management on your PC—making things work more efficiently. **Defragment** your hard drive (this reassigns file usage on your hard drive, cleaning up all of the fragments of files that are left after saving and deleting files), and delete files you no longer need. There should be an icon on the desktop for your Recycle Bin (in Microsoft operating systems—it can also be found on the control panel in Windows Explorer) or Trash (in Apple operating systems).

Inoculate against and kill viruses and spyware. Keeping your antivirus up to date and using regularly updated anti-spyware programs will save you from having a slow and unsafe PC. (See page 33.)

Spyware is monitoring software that tracks your actions on the Internet. These cannot only violate your privacy, it also uses your computer's resources for the benefit of the people tracking you. **Viruses** and **malware** are programs that are designed to slow or damage your computer, steal your information, or attack other people's computers. If you don't keep your computer protected or remove these programs quickly, it can damage your files, your programs, and even your computer.

Installing antivirus and anti-spyware programs is not enough—you need to keep them up-to-date. An ounce of prevention is called for here as these programs can be extremely difficult to remove. If your computer does become infected, the only option may be to reinstall your operating system and all your programs, which also means losing any files that haven't been backed up.

Before You Call—Simple Fixes to Try First

The most common reason something goes wrong on a PC is newly installed software or hardware. If you install a new program, and it does not work right or causes other things to not work right, try going to the software maker's website and see if they have a patch or an update to the program.

If the new program causes the PC to stop working, restart the PC, hit the F8 key, and choose **Safe Mode** from the list. It will take a lot longer to start and the screen will look different. When it does finish loading, go to the control panel and choose **Add/Remove Programs**. Remove the newly installed program and then restart your computer. This usually fixes the problem.

As for new hardware, you can try the same thing. Check the manufacturer's website and check for updates. If there aren't any, remove the hardware and restart.

Preparing to Call for Help

Before you call tech support for your malfunctioning device, it always helps to *get your story together.*

- Be calm.
- Be sure you have time to address the problem before you call. You will sit on hold, sometimes for quite some time. It's best to be on a cordless phone. Plan to have a diversion on hand to entertain yourself while listening to endless "you're call is very important to us" messages.
- Have a notepad to take notes. Get the customer service representative's name and ID number—oftentimes, you will be given a case number, trouble ticket number, or service-call number.
- Be patient, they have a script they have to follow to eliminate common problems. But if you are calling back on a previously discussed item, make sure that you give them the previous call's reference number.
- If they want you to perform a process and then call back, tell them that you would prefer that they hold on the line—be firm but polite.
- Escalate to a higher-level phone representative or technician when you feel that you are not making progress with the initial customer-service agent.
- If someone has a heavy accent, and you truly have a hard time understanding them or believe that they do not understand you, politely ask to speak with a supervisor. We all have accents, and they may also be having trouble understanding yours, but as their customer, you should be able to find someone with whom you can work.
- If you don't understand an instruction, ask them to explain it more clearly to you.

- If they are unable to fix the problem, or give you advice to remedy it yourself, ask them to let you know of any other options available.
- If you are not satisfied that the customer-service department's service, write a letter to the manufacturer explaining your dissatisfaction and ask for some resolution of the problem.

Helping Them, Help You

Now that you've prepared for contact, here's how you can *help them help you.*

- Let them know how the problem started.
- When errors pop up, read the message and jot the error message down. Often you can research the error yourself, the resulting answer might be overly technical, but the technician may know what direction to look towards to address the problem.
- Tell them what you have already tried.
- Be sure to call the right person first—is your Internet connection dropping? Call your Internet provider first, before calling your hardware manufacturer. If you are told that the network is functioning properly, then try the hardware manufacturer.
- Follow instructions carefully, don't jump ahead.
- Know your model number and serial number—you should have the manual with the receipt stapled to it.

Bringing in the Pros and Avoiding Rip Offs

Until you are much more comfortable with how your

PC works, there are some times when you will need a pro to help out.

What to ask when hiring someone to fix your PC:

- *Do you charge a diagnostic fee, and does the fee go toward repair if I choose to use your services?* A small fee is normal as it can take some time to figure out what is exactly wrong and how much time it will take to fix, but as finding the problem is more than half the battle on these things, they should apply the fee to the repair.

- *Will they come to your house to do the repair, and is there an extra charge for this service?* There is most likely an extra charge, but there are many reasons to have your computer fixed at home. One is convenience, of course. It is sometimes difficult to disconnect your device and bring it to their repair facility. But one thing you may not have considered is that with an on-site service call, you can watch them do the repair and check that they are not looking at files they don't need to be looking at (see the next bullet point). It also means that the repair will be the top priority to the technician while they are at your site, instead of landing on an ever growing to-do list at the shop.

- *What is your policy on keeping my data private?* This is a big issue—you don't want someone snooping around your personal files, and the people you hire should have a clearly stated policy on this. Ask to see it if possible.

- *Do you perform the repairs here, or do you ship them somewhere else to do the repair?* Most places should be able to do the repair in-house. If they tell you they have to ship it out to do the repair, consider going somewhere else as this is extremely time consuming and there are many horror stories of systems never coming back. Ask them what coverage they have for lost, damaged items and get it in writing.

- *Is there a warranty on the repair?* Be sure that there is at least a thirty-day warranty on the repair.

Working with the Technician

As mentioned above when calling for customer service, it is important that you:

- Clearly state the problem, and how it came about.
- Understand what they are describing as their diagnosis.
- Understand the options you have to repair it.
- Are not afraid to walk away from a situation in which you are uncomfortable.

Things to Look Out For

Many shops will try to convince you that you need a whole new PC. This is rarely the case. If they come back to you and say you need to replace the whole thing, pay the diagnostic fee and get a second opinion. Ask for a detailed reason as to why they think the whole thing needs to be replaced. The most common reason is it will cost more to fix than repair, and as we have said, this is almost never the case unless it is a laptop monitor.

Expensive upgrades—sometimes you will be told that something is not fast enough or big enough and needs to be upgraded or that you need to buy this special software to fix the problem. Most likely they don't know what the problem is or how to fix it—again, pay the diagnostic fee and get a second opinion. Take your business elsewhere.

They tell you that it's going to take several weeks to fix. Ask why this is. Usually this is a sign that they don't have the staff to fix the problem, or there is another problem in the shop. A long time frame increases the likelihood they will lose your system or forget about you.

We have found that asking around to see what services other people have used is a good idea. If you know an IT person, ask them if they can recommend someone or someplace. They are most likely to know the good shops or technicians in town.

Warranties

For portable electronics—phones, media players, and gaming systems—you will often only have the warranty repair as an option for prices less than replacement.

- Know your warranty period, have the proper documents and your serial number.
- You need to follow their directions exactly.
- Always use traceable mailing/shipping methods. Pay extra for insurance and tracking if using the postal system.
- Get a repair ticket number, and follow up BEFORE the promised return date, again as with all tech-support calls, take notes and keep tracking numbers together.
- If your product is not returned when promised, follow up immediately and ask to speak to the manager.
- If your product is out of the warranty period, there are other options than simply replacing your beloved (if it is beloved!) digital device. As local repair shops are sometimes reluctant to work on these sometimes highly

specialized devices, try using one of these services recommended on the Consumerist.com website, a consumer advice and information blog:

www.techrestore
www.iresq.com
www.llamma.com
www.xboxrepairguide.com
www.weaknees.com

Checklists

Preparing Before Things Go Wrong

Do you have a place where you keep all of your manuals, warranties, installation disks, receipts, etc.? If so, where is it? ______________________________

Have you registered your devices/software? If so, which ones?

____

____

Have you claimed all of your rebates? ______________

Do you have scheduled maintenance tasks (checking for updates, defragmenting your hard drive, deleting unused files, etc.) on your computer?__________________

Do you have a backup regimen for your computer? What is it? How often? ______________________

Have you asked friends or your company's IT department for recommendations on preferred repair shops/services? If so, provide the list: ________________

____

____

Do you know the warranty term for your major devices (computer, printer, camera, etc.)? List them:

____

____

____

____

Trouble Call Reference Sheet

Before you call:

Nature of problem:

____

____

When did it start?

____

____

What have you tried?

____

____

And what were the results?

____

____

Manufacturer/Model and serial number:

____

Record of the call:

Who did you call?

____

Which phone number did you use?

____

Customer-service representative name and/or ID number:

____

What fixes were suggested?

____

____

Did this work?

____

____

Were you given a ticket or trouble-order reference?

____

If the suggested fix didn't work, did they give you suggestions for follow up or the next step?

____

____

For on-site service, did they give a time frame and any possible costs?

____

____

For off-site service, what were the shipping instructions?

____

____

Address shipped to:

____

Tracking number of shipment:

____

Date shipped:

____

What number were you given for follow up?

____

Was the item repaired and returned? (If not, call back and reference ticket number.)

Trouble Call Reference Sheet

Before you call:

Nature of problem:

____

____

When did it start?

____

____

What have you tried?

____

____

And what were the results?

____

____

Manufacturer/Model and serial number:

____

Record of the call:

Who did you call?

____

Which phone number did you use?

____

Customer-service representative name and/or ID number:

____

What fixes were suggested?

____

____

Did this work?

____

____

Were you given a ticket or trouble-order reference?

____

If the suggested fix didn't work, did they give you suggestions for follow up or the next step?

____

____

For on-site service, did they give a time frame and any possible costs?

____

____

For off-site service, what were the shipping instructions?

____

____

Address shipped to:

____

Tracking number of shipment:

____

Date shipped:

____

What number were you given for follow up?

____

Was the item repaired and returned? (If not, call back and reference ticket number.)

Trouble Call Reference Sheet

Before you call:

Nature of problem:

____

____

When did it start?

____

____

What have you tried?

____

____

And what were the results?

____

____

Manufacturer/Model and serial number:

____

Record of the call:

Who did you call?

____

Which phone number did you use?

____

Customer-service representative name and/or ID number:

____

What fixes were suggested?

____

____

Did this work?

____

____

Were you given a ticket or trouble-order reference?

____

If the suggested fix didn't work, did they give you suggestions for follow up or the next step?

____

____

For on-site service, did they give a time frame and any possible costs?

____

____

For off-site service, what were the shipping instructions?

____

____

Address shipped to:

____

Tracking number of shipment:

____

Date shipped:

____

What number were you given for follow up?

____

Was the item repaired and returned? (If not, call back and reference ticket number.)

Glossary

Advanced Audio Coding (AAC): An audio media format; it is the default media setting used by iTunes and other Apple products, Sony PS3, and others. It may have the following filename extensions (the suffix after the filename) .m4a, .m4b, .m4p, .m4v, .m4r, .3gp, .mp4, and .aac.

Amazon Kindle: An e-book reader sold by and integrated with Amazon.com.

Analog: As opposed to digital; a system where there is a continuously variable signal or wave.

Analog phone: As opposed to a digital phone; a phone based on a nondigital phone signal.

Anti-spyware software: Software that detects and can remove software that provides the unauthorized monitoring of a user's computer and/or Internet use.

Anti-virus software: Software that detects and protects your computer from malicious software called viruses.

ATSC (Advanced Television Systems Committee): This is a set of standards for digital television signals to replace the analog NTSC (National Television System Committee) in North America.

Automatic White Balance (AWB): A feature on a digital still or video camera that adjusts the picture's color temperature so that the color white is faithfully reproduced.

Backup: A copy of or the act of copying one's data onto another source to recover it if there is some sort of failure.

Bandwidth: The capacity for transfer of information.

Bit rate or sample rate: The number of bits recorded in a period of time; usually given in bits per second: for example, Mbps (megabits per second) or Gbps (gigabits per second).

BlackBerry: A brand of device that uses a wireless e-mail system, usually with a phone and other PDA (personal digital assistant) tools.

Blog: Shortened from "web log"; a blog serves as an online diary or personal website meant to be shared with other Internet users.

Blu-Ray: An optical storage format that is primarily used for High Definition video and data storage. Blu-Ray disks are the same physical size as a DVD or CD.

Bluetooth: A short-range wireless transmission technology that creates a Personal Area Network (PAN) so that different devices can communicate.

Broadband: An Internet connection that has a much larger data capacity than dial-up.

Browser: The software that enables you to view the Internet; some examples are Microsoft's Internet Explorer, Apple's Safari, or Mozilla's Firefox.

Burn or burning: Writing or copying data onto a disk; for example, burning a CD or DVD.

CableCARD: A plug-in card that allows U.S. consumers to receive digital video signals from cable providers.

Car charger: A device that recharges batteries using the power of your car's electrical system by connecting it through special outlets in your car.

Cathode Ray Tube: Refers to a type of display or monitor.

CCD: See Charge-Coupled Device.

CD: See Compact Disk.

CD drive: A hardware device that reads CDs.

CD writer: A hardware device that stores data on CDs.

CDMA: See Code Division Multiple Access.

Cell sites or cell towers: The points of transmission of cellular phone signals.

Cellular or cell phone: A wireless phone that doesn't use telephone lines, but instead uses a network of phone reception cells.

Central Processing Unit (CPU): The part of the computer that does the basic work of the computer: processing data and controlling the other components.

Charge-Coupled Device (CCD): An electronic light sensor used in some digital cameras.

Closed captioning: A text display of the audio portion of an audio-visual broadcast.

CMOS: See Complementary Metal-Oxide Semiconductor.

Code Division Multiple Access (CDMA): A type of cellular phone network, primarily used in North America by Verizon Wireless, Sprint, and some smaller providers.

Color temperature: The concept that describes the relative appearance of colors in varied lighting environments.

Compact disk: A type of optical storage media.

CompactFlash: A storage device format that uses flash memory.

Complementary Metal-Oxide Semiconductor (CMOS): An electronic light sensor used in some digital cameras.

Content Scramble System (CSS): A type of copy protection most often used on DVDs.

Copy-protected or copy protection: A process used by the manufacturers of digital media or software to prevent unauthorized copying or use; sometimes referred to as Digital Rights Management (DRM).

Cordless phone: A phone that communicates wirelessly to a base station that is connected to a telephone line.

CPU: See Central Processing Unit.

CRT: See Cathode Ray Tube.

CSS: See Content Scramble System.

Data plan: An added service on a cellular phone contract that allows the user to access the Internet.

Dead zones: Areas where there is no cellular phone service on a particular network.

DECT: See Digital Enhanced Cordless Telecommunications or Telephone.

Defrag or defragment: A process that collects fragmented data on a computer's hard disk to make file management more efficient.

Depth of field: In photography, the focal point of a scene that appears sharp relative to objects in the foreground or background.

Dial-up: A type of Internet connection that uses the telephone lines and a modem.

Digital: A signal composed of binary digits as opposed to analog waves.

Digital camera: A camera that stores images as computer memory as opposed to storing on film.

Digital compression:The process of encoding information in fewer bits by removing repetitive or extremely similar information.

Digital Enhanced Cordless Telecommunications or Telephone (DECT): A digital standard for cordless phones that differentiates its signals from other electronic transmissions.

Digital phone: A phone that uses digital signals as opposed to phones that operate over analog phone transmission lines.

Digital Rear Projection Television: A television that uses digital technology to create a television image that is projected internally onto a screen.

Digital Rights Management (DRM): See Copy Protection.

Digital Single Lens Reflex (DSLR) Camera: A digital camera that uses an automatic mirror system with a mirror or prism that reflects light onto the sensor to capture images digitally.

Digital Subscriber Line (DSL): A technology that allows for the digital transmission of information over the telephone lines; usually referring to an Internet connection and voice-call transmission over the same telephone line.

Digital Video (DV): A digital video format.

Digital Video Disk (DVD): A type of optical storage media.

Digital Video Recorder: A video recorder that captures the video and sound images digitally and stores it on digital media. It also refers specifically to a device that records programming for time-shifted viewing and rewind features on live television viewing.

Digital Visual Interface (DVI): A digital video interface that allows for efficient transmission of video images between devices.

Disk Operating System (DOS): The basic operating system that manages the storage devices and the information stored on them.

Docking station: A device that connects a portable device with a fixed device, usually a computer.

DOS: See Disk Operating System.

Download: A file transferred from a computer to another device or storage media.

DRM: See Digital Rights Management.

DSL: See Digital Subscriber Line.

DSLR: See Digital Single Lens Reflex Camera.

DTV: Digital Television .

DTV transition: The transition from analog television broadcasting to all-digital free television broadcasting scheduled for February 2009 in the United States.

Dual core processor: A CPU that uses two integrated processors that act as one single integrated circuit.

DV: See Digital Video.

DVD drive: A hardware device that reads DVDs.

DVD player: A device that plays DVDs and sends out an audio and video signal to a monitor or television.

DVD writer: A hardware device that stores data on DVDs.

DVI: See Digital Visual Interface.

DVR: See Digital Video Recorder.

e911 or E911: see Enhanced 911.

E-book: Short for electronic book; a digital file of a book that can be read on a computer or e-book reader.

Encoding: The process of converting a digital signal from one format to another.

Enhanced 911 (e911 or E911): A system in North America that automatically connects and displays a physical address with a caller into the 911 call center.

External hard drive or disk: A standalone device that stores data outside of the computer's internal hard drive.

FCC: See Federal Communications Commission.

Federal Communications Commission: The government regulatory agency that regulates and enforces the federal laws and rules for the fields of telecommunications and broadcasting.

FHSS: See Frequency Hopping Spread Spectrum

Fiber Optic Service (FiOS): A fiber optic television, broadband Internet, and voice service sold by Verizon.

Filename extension: A suffix attached to a file name when stored to digital media that indicates the file format under which it was saved; some examples are .doc, .jpg or .mp3.

FiOS: See Fiber Optic Service.

Firewall: A device or software program that is designed to keep out potentially harmful Internet traffic when placed between an Internet connection and a router or device.

Firmware: The software that is embedded on a piece of hardware that allows it to function.

Flash Memory: A type of computer memory that doesn't require the computer to be turned on to maintain the information (referred to as non-volatile memory).

Focal length: The focal length of a lens determines the amount it magnifies distant objects.

Frequency Hopping Spread Spectrum (FHSS): A method of radio transmission that spreads a signal among different frequencies making it very hard to interfere with or intercept.

Gamer: A term used to describe an avid videogame player. Gamers have their own culture and even language—1337 or *Leet*—which combines letters, symbols, and gaming neologisms.

GHz: See Hertz.

GIMP: A free software suite that is used for graphic design.

Global phone: A phone that can be used both domestically and abroad because it can operate under different bands or networks.

Global Positioning System (GPS): A satellite-based technology that provides geographic location and directions.

Global System for Mobile Communications (GSM): A type of cellular phone network, the most popular standard for mobile communications in the world. In the U.S., T-Mobile and AT&T are the two largest GSM network providers.

GPS: See Global Positioning System.

Graphical User Interface (GUI): The software program that uses graphic elements to provide ease of use to computer users.

GSM: See Global System for Mobile Communications.

GUI: See Graphical User Interface.

Hard Drive or Disk: A device used that serves as the main storage of data on a computer. It stores data magnetically on a rotating disk. It is non-volatile storage, meaning it doesn't need to be turned on to retain the information.

Hardware: The computer or electronic equipment itself, as opposed to software, which is a set of instructions written in program language.

HD or High-Def television: High Definition, usually referring to video signals higher than standard definition (480i in North America).

HD-Ready or HD monitor: Video monitors that can display HD content when connected with HD input but does not have a built-in HD tuner.

Hertz (Hz): A unit of frequency; in computing used to indicate the speed of a CPU, usually in Gigahertz (GHz).

High-Def: See HD or High-Def Television.

High-Definition Multimedia Interface (HDMI): Is an interface that can transfer both HD video and audio components of a signal.

HMDI: See High-Definition Multimedia Interface.

Hz: See Hertz.

ICE: See "In Case of Emergency."

Image sensor: The processor on a digital camera that converts the picture to digital information.

In Case of Emergency (ICE): A popular solution to storing an emergency contact name and number in a cell phone; emergency contacts are listed under the acronym "ICE," thereby giving emergency personnel a standard place to look for that information.

Internet: A worldwide series of interconnected networks that communicates using Internet Protocol to transfer data such as e-mail, file transfers, and chat. The graphical component of this is called the World Wide Web or simply the Web.

Internet Service Provider (ISP): A company that provides access to the Internet and its resources, such as e-mail, by connecting you to their network servers via dial-up or broadband methods, such as cable, DSL, or fiber optic lines.

iPod: A brand of Personal Media Player produced by Apple.

ISP: See Internet Service Provider.

iTunes: A digital music and video store operated by Apple.

Key: In software, this is the individual code assigned to a particular copy of a program that establishes it as a legal and useable copy as part of a copy-protection scheme.

Landline: Refers to a standard telephone line as opposed to a mobile phone or digital phone service.

LCD: See Liquid Crystal Display.

LED: See Light Emitting Diode.

Light Emitting Diode (LED): A semiconductor light that emits light when electricity is applied; used in basic displays such as on/off indicator lights.

Liquid Crystal Display (LCD): A thin video display made by projecting light through a set of color or monochromatic pixels.

Logicboard: See Motherboard.

Long-distance provider: A company that connects your local phone service to the network of other domestic and international phone systems.

Lux (Lx): A measurement of light or illumination.

Lx: See Lux.

Mac or Macintosh: A computer made by Apple.

Malware: A term used to describe computer programs (software) that are intentionally damaging or meant to divulge information without the user's knowledge or permission.

Media PC: A personal computer that also serves as a hub for your entertainment or other media; usually there is a broadcast capability to other devices in your home network.

Megapixels: see Pixel.

Memory: A device or component that can store data from a computer or other digital device.

Memory card: A flash memory storage device available in different formats such as a PC card, SD, Memory Stick, or Multimedia card.

Memory card reader: A device that can upload data from a memory card.

Memory card writer: A device that can download data onto a memory card.

MHz: See Hertz.

Migrate: In electronics, moving from one device or technology to another.

MiniDV: A video format.

Mobile phone: See Cellular Phones, the term Mobile Phone is more commonly used outside of the U.S.

Mobile Internet or web browsing: Using your cell phone to access the Internet.

Monitor: A device that displays video images.

Motherboard: The main circuit board of a computer, it holds the CPU, memory, component connectors, and other essential components of a computer.

Motion control: A technology used in video gaming that detects and controls movement through the position and motions of the controller instead of particular buttons or keyboard commands; first popularized in gaming by Nintendo's Wii system.

Mouse: A computer peripheral that acts as a pointer device when moved along a flat surface and clicked.

MP3 Player: See Portable Media Player (PMP).

Multi-Core processor: A CPU that uses more than one integrated processor, but which acts as one integrated circuit.

Pixel: One unit of display; usually refers to a large quantity that together composes a video or still image; commonly referred to in units of megapixels.

Plasma: When used to describe a monitor type, it is a thin video display made by exciting cells of inert gases contained within two pieces of glass.

RAM: See Random Access Memory.

Random access memory: Memory that can be accessed and stored dynamically by a computer during operation.

Sample rate: See Bit Rate.

Software: A set of instructions written in program language that operates hardware devices.

Standard definition: Refers to a video signal broadcast below High Definition, standard definition in North America broadcasts up to 480i.

Upload: A file onto a computer from another device or storage.

Virus: A malicious software program that tries to replicate itself by copying itself onto your stored data or transmitting to other users via e-mail or other file transfers.

Web or World Wide Web: The graphical component of the Internet, which consists of web pages or web sites.

Helpful Websites

Now that you are less Digitally Daunted, here are some websites that can give answers, offer advice, and introduce you to new products or ideas.

DigitallyDaunted.com

Visit us at this book's companion website to see updated information and product reviews and to get answers to your questions.

Manufacturers' Websites

Microsoft.com—While this huge site may seem overwhelming at first, you will be glad to see all of the resources available there. After "landing" at their homepage, look for the particular Microsoft product you are using and click on the menu to find all sorts of information. You can bookmark the web page for the particular products you use (such as Windows, Microsoft Office, Xbox) and visit the site for updates, articles, and technical help.

If you have any error messages appearing on your computer, you can search the error name or number and possibly find an easy solution; at least searching for the error message will give you some extra information before you seek repairs. There are also how-to videos for using various Microsoft Office programs such as Excel and Outlook.

Apple.com—Just as with Microsoft's site, there is a wealth of information on Apple products and programs. Again, simply search for your product to find more information about how it works, check for updates, and get help with technical problems.

On the Mac portion of the Apple website, they have video tutorials and articles about how to make it all work together.

Your computer manufacturer's site (Dell.com, HP.com, Acer.com, Lenovo.com, Toshiba.com, and others):

Each of the major manufacturers has a "support" section and most offer a variety of information and tools for using their computer models and programs.

The manufacturer's site for each of your gadgets: If you have questions about any of your gadgets, if you've lost your manual, or if you are having technical problems, you may want to check out the manufacturer's website. We have found that there is often a wealth of information out there even for the most inexpensive device. This is not always the case, but it is a logical first step when you need information.

Government Resources

DTV2009.gov—This is the official government website with information about the upcoming conversion from analog to digital television, including the coupon program for buying an analog-to-digital converter box.

Fcc.gov/cgb/—This is the Federal Communications Commission's Consumer and Government Affairs Bureau's website. It includes information on a variety of topics, including: the Do Not Call list, junk faxes, closed-captioning, and much more. There is also a section for Consumer Alerts.

Ftc.gov/bcp/consumer.shtm—The Federal Trade Commission's Consumer Information site has information about unfair trade practices, how to protect yourself against scams, and how to be a smart consumer.

Online Tools and Information

Google.com—Besides the Internet search that you already know, did you also know that Google has office software, photo storage, and blogging software? After you go to the main page, just hit "More," and you can see all of the free tools and search options.

Wikimedia Foundation projects—
http://wikimediafoundation.org/wiki/Our_projects
You may be familiar with Wikipedia, which is a free online, collaborative encyclopedia, but the nonprofit Wikimedia Foundation also has many more tools, including a free dictionary, thesaurus, collection of quotations, access to free textbooks, and a collection of images and other media that are subject to what is called a Creative Commons license. Basically, Creative Commons licensed content is free to use and share subject to certain restrictions, such as the requirement that appropriate credit is given.

Wikia.com—Wikia is a related, but commercial, advertisement-supported collection of collaborative websites, including travel information, sports communities, entertainment forums, and other user-created sites.

Consumer Technology Information

Cnet.com and Download.com—This is a commercial site that has reviews, information, and news about consumer technology. This also includes Download.com, which provides secure downloads of software, including a wealth of free software, accompanied by editorial and user reviews.

Charitynavigator.org—Charity Navigator provides the tools necessary to judge a charity that you are considering contributing to and advice on choosing a charity wisely.

BBB.org—The Better Business Bureau is the place to go to see if a business has consumer complaints, to report a consumer complaint, or to seek help in resolving a complaint.

Your individual state's Consumer Affairs or Consumer Protection Agency

You can search at your state's main government page for the agency that handles consumer regulation and information. They usually have a frequently asked questions section and provide resources to resolve consumer complaints.

Consumerist.com—The Consumerist is a top-rated blog devoted to discussing consumer affairs from individual stories, news, and tips on getting better service to resolving conflicts. While the blog's individual stories (usually links to other news sources) are interesting, sometimes the best information can be found in the comments.

Photo Sites

These are a few of the most popular photography sites: Kodak.com, Shutterfly.com, Picnik.com, Photobucket.com, FotoFlexer.com, Picasa.com, and Photo.net.

Our Favorite "Just for Fun" Sites

LoC.gov—The Library of Congress—you could get lost for years browsing through here!

Blogs.usatoday.com/ondeadline—A clean and mean breaking news site from *USA Today*.

Nasa.gov—The space agency's site has information, news, and striking photography.

BoingBoing.net—Sometimes not for the faint of heart or easily offended; the subtitle of this blog is "A Directory of Wonderful Things."

IMDB.com—The International Movie Database: For the casual movie fan and the absolute cinephile; a compendium of everything film, actor, and television.

Allmusic.com—Find out everything you want to know about your favorite bands, who sang that song, and what artists you might like based on artists that you already know.

Green Tips

Electronics themselves and the process of manufacturing them can be environmentally detrimental. You can limit these negative impacts in several ways: buying as green as possible, donating reusable products, recycling and disposing of your products responsibly, and using power efficiently.

Buying Green

Look for electronics bearing the Energy Star label. This means that they meet or exceed the U.S. Environmental Protection Agency's most stringent energy-efficient requirements. Research the manufacturer's "green" policies. Do they take steps to reduce the use of toxic chemicals? Do they use recycled materials? Do they use efficient packaging?

Donating Reusable Products

If you have relatively modern and functional electronic equipment, there are many nonprofit organizations that would want your used equipment. Contact your local school system, library, or local government's volunteer office for these opportunities. Be sure that you clear all identifying and personal information from your equipment to protect your privacy and your identity. Be sure you remove all files, e-mails, e-mail contacts, and transferable software.

There are utilities (software) that can do this for you—see sites such as CNET.com or Techsoup.com for lists of software that can "clean" your computer. If you are donating other used electronics, see your manual or visit the manufacturer's website to see how to clear your data from the device.

Recycling and Proper Disposal

You can contact your old computer's manufacturer or your new computer's manufacturer for information about recycling programs they may have in place. Cell phone providers have all started to provide their consumers with ways to recycle used cell phones. You can also check with your local government's waste management office for local recycling options. If you still cannot find a responsible way to dispose of or recycle your electronic equipment, you can visit the Consumer Electronics Associations recycling website at mygreenelectronics.org. They have a tool that is product specific with suggestions on where and how to recycle. Remember, almost every consumer electronic device has a battery that is toxic waste. Follow your local guidelines in disposing of anything that you do not donate or recycle.

Using Power Efficiently

All electronic devices use electricity. On battery-powered devices, there's usually a gauge to show you that the battery power is depleting. It's harder to see the so-called "phantom power use" of your plugged in appliances. But why use that power when you don't need it? It's both costing you money and causing more pollution. You should turn off your computer when you are not using it and consider unplugging large electronics such as your television that draw power even when not on. One idea is to put your home electronics on power strips that you can switch off before going to bed or going on vacation.

Acknowledgments

The authors would like to thank Amy Fries, our editor at Capital Books, for her enthusiasm, patience, hard work, and insights. We thank Kathleen Dyson and Michie Shaw for their work on the book's production, Jane Graf for her marketing insights, Kathleen Dyson for the cover design, and Kathleen Hughes for her faith in the book and its authors.

We take this opportunity to thank our families for their support and to our friends and co-workers who have shared their good wishes and encouragement.

About the Authors

Sean Westcott is an IT professional dating back to before the World Wide Web. He was in his high school AV club where he explained technology to his teachers, and ever since he has made it his specialty to bridge the knowledge gap between geeks and the rest of the world.

Jean Riescher Westcott took a History of Computers class back in 1985 and spent a summer studying programming in high school, but has been more of a book geek since she finished her law degree. She spent a long, wonderful time in independent bookselling and now works in book publishing. Sean and Jean have two children and live in Northern Virginia.